Science, Math and God

Martin Spiller

Science, Math & God

Introduction

Is life after death a real possibility? From the date of publication of Isaac Newton's Principia in 1687 until the early years of the twentieth century, science answered with a resounding "NO". Newton's brand of physics simply left no room for a spiritual world. However, things changed at the beginning of the twentieth century, and physics no longer can shut the door on the existence of a spiritual world.

It turns out that Sir Isaac Newton doesn't run the show anymore. The circus of reality now belongs to Albert Einstein, Erwin Schrodinger, Neils Bohr and Werner Heisenberg. These scientists murdered materialism and provided us with a foothold for God!

This book does not try to prove that God exists. It is simply lays out the history and current state of science and demonstrates that not only does science NOT explain everything. It simply cannot and never will be able to explain everything. The fundamental questions about our existence on earth will remain opaque to scientific inquiry forever!

This book was written for people who sense that there must be more to life than constantly striving for

1

fame, fortune and sex. They want to believe in God but find it difficult to commit because of the largely materialist society in which they live and the constant reminders from those around them that science explains everything, making God and his Heaven a relic of a past superstitious age. This belief is simply no longer true!

Everything you will read here is based on hard science. There's no religious dogma here. It details the history of modern science as well as the dominant theories that touch on the nature of our reality and the question of creation. NO! This book does *not* justify "Creation Science"! But it does explain in depth just why modern cosmology's supposedly scientific creation theory is no less mythical than the Old Testament version of Genesis.

People in our digital age often discover that an entirely material world is not an entirely happy one. Younger people, especially those under the age of thirty may not believe this statement, but as these people age (as they all presumably want to do), the ordinary challenges of their lives begin to wear on them, and even the most skeptical among them begin to have doubts.

The purpose of this book is not to prove the existence of a spiritual realm, but, rather to allow you to walk up to the boundary of our reality and peer over the edge. You can't believe in God without first understanding

that He, as well as all of His subordinate spiritual
entities have a place to live.

1

Classical Physics and the Clockwork Universe

Do you really believe that the moon isn't there when nobody looks?
Albert Einstein

Prior to 1905, materialists could look to Sir Isaac Newton for a scientific confirmation of their beliefs. Newton's laws were fairly straightforward and left no room for God or his spiritual realm. Sir Isaac explained the relationship between mass, motion and gravity using mathematics. Newton invented what we today call "classical physics".

In classical physics, any given future state of the physical world is completely determined by its state at an earlier time. Think of a billiard table with all the balls racked up in an equilateral triangle, sometimes called a "pyramid". The pool player uses his cue to aim the white cue ball at the pyramid. After the cue ball strikes the pyramid, all the balls roll off in seemingly random directions. They bounce off each other and off the walls of the pool table and scatter all over the table.

Now, imagine, if you will, that this is a magic billiard table. This table is absolutely frictionless and allows the balls to keep rolling and bouncing without ever stopping. The balls simply continue to roll and ricochet off each other and the bumpers on the sides of the table forever. One ball knocks into another one imparting new momentum to both balls which keep rolling in different directions and at different speeds in a never ending dance.

Classical (Newtonian) physics assumed that the entire universe was constructed of atoms that resembled the billiard balls in the above example. They believed that these atoms all engaged in a never ending dance of collisions beginning when the universe was first created and continuing to the present and beyond. One atom strikes another, each strike imparts a new momentum to both atoms, and these, in turn, strike other atoms in a never ending cascade of events. The changing momentum and position of each atom could technically be calculated if one had a super computer that could keep track of every strike and trajectory of every atom in the universe. The computer could thus predict the continuing evolution of the universe on a microsecond by microsecond basis.

In other words, according to classical physics, the complete history of the physical world was determined for all time when the universe first came into existence. Thus everything in the universe advances from one second to the next, each new second determined by the state of the atoms in the second before. After Newton, scientists, as well as the elites of the day came to believe that if an omniscient observer could know the mass, position and velocity of every particle in the cosmos at its first moment, he could calculate the future with perfect certainty. Thus, Newtonian physics saw the universe like a gigantic *clockwork*, each gear turning against other gears, each movement pre-determined by the movements of all the other movements working in unison.

The exact conditions at the beginning of the universe set the clock going and everything that would ever happen any place in the universe, including the very thoughts in your mind was destined to happen from the very beginning of time. This idea of a clockwork universe set the tone for the doctrine of **predestination**.

A Scottish philosopher named David Hume (1711 – 1776) further refined the philosophical implications of Newton's classical physics rejecting the overarching religious thinking of his day and replacing it with the modern concepts of **empiricism** (observation and experimental evidence), **skepticism** (of established ideas, but especially of religious and spiritual ideas) and **naturalism** (the belief that only natural laws and forces operate in the world). Essentially, Hume simultaneously invented the philosophical underpinnings of both modern science and modern atheism. Today, his philosophy is called *Secular Humanism* - named appropriately after Hume himself. Secular Humanism was soon embraced by the intellectuals and elites of his day, and remains the de facto religion of modern elites as well as millions of ordinary people who hold materialistic values. We will be examining secular humanism and Hume in more detail later.

The philosophical ramifications of the new physics were very upsetting to the prevailing views of everyday life. Prior to Newton, everyone could experience, at the very least, the comfort of a belief in God and a place in Heaven after an impoverished and often painful life. But the new physics allowed no place for a god. Indeed, there was no place for a heaven. And the elites of the day, as they have done in every age since, have tried their best to stamp out all those "superstitious" beliefs of God and spirits; beliefs that humans have taken comfort in since the beginning of time. If you are reading this, you already know that in over three hundred years of trying, they have not succeeded.

Predestination and its philosophical stepchild, secular humanism assume that there is no such thing as free will. It assumes that everything you have become, and everything you will ever become, everything you think and everything you ever will think was predestined to happen eons before you were born. With an ultimately powerful computer, one could not only predict the future, but one could also retrodict the past by working backwards. But time always progresses forward, and as the universal clock unwinds, so does your entire life. The universe, your life and all earthly circumstances are predestined to progress to an ultimate end, in the grand arc of history and there is nothing you can do can do to change it. In classical physics and the philosophies it spawned, there is no such thing as **free will**! You only THINK you have free will, but even that thought was predestined at the dawn of the universe.

Reaching the limits of Classical physics

Sir Isaac Newton published his *Principia* (his "theory of everything") in 1728. The *Principia* laid out all the laws of motion that were understood up until that time, as well as introducing the mathematical concept of gravity. Newton's laws understandably assumed that both time and mass were constants. In addition, he believed that light was made of tiny particles (which he called "corpuscles") and which were subject to all the laws that any physical objects had to obey. Thus, his theory of light predicted that the speed of the light corpuscles would vary depending on the frame of reference of the light source. The terms "frame of reference" and "relativity" need to be defined.

If a car is moving toward you at 50 miles per hour, and you are walking toward it at 4 miles per hour, then your *relative* speed toward the car is 54 mph. On the other hand, if you are walking away from the car, it would be approaching you at only 46 mph. In other words, your speed with respect to the car changes depending on your *relative* motion with respect to the car. In this context, you and the car are isolated from everything else in your own "frame of reference". Likewise, Newton believed that if you were moving toward the source of the light, then within your own frame of reference, the speed of the light particles should appear to be faster than if you were moving away from the source. What made light seem different from ordinary matter was its enormous speed which was first measured 52 years before Newton published his Principia by Danish astronomer Ole Römer in 1676.

But over the next 150 years, science had progressed, and by 1887, as a result of numerous experimental observations, scientists had decided that light behaved more like it was composed of waves of energy instead of discreet particles. Up until that time, it was assumed that waves have to travel in some sort of compressible medium. Ocean waves travel in water, and sound waves travel in the air. If light was composed of energy waves, scientists of the day assumed that it too must require some sort of medium to propagate. Furthermore, this medium had to fill the space between the stars. Otherwise the light from those stars could never reach us.

They called this medium the "luminiferous aether" (pronounced "ether"). This wasn't just a one-off screwball idea. It was an absolute essential because Newtonian physics was understood to apply to waves of energy as well as to particles of matter. It was also known that the speed of wave propagation depends upon the density and motion of the medium. While the properties of the medium carrying light waves were unknown, it was widely assumed that this medium MUST exist, and if scientists could discover that light traveled at different speeds depending on the medium's density and direction of flow, then other properties of the "luminiferous aether" itself could be calculated.

Albert Michelson and Edward Morley were scientists working at what is now Case Western Reserve University in Cleveland, Ohio. They reasoned that if the luminiferous aether exists in outer space, then the earth must be moving through it like a fish through water. Therefore, there ought to be a measurable difference in the speed of light when measured at different angles with respect to the motion of the earth in its orbit around the sun. They therefore ran an experiment comparing the speed of light in perpendicular directions. They were surprised when they could discover no variation in the speed of light regardless of the angle used to measure it.

They concluded from this that the luminiferous aether does not exist and that light could not be composed of classical particles either, since it does not seem to obey any of the relativistic laws known to physics up until that time. If light was composed of waves propagating through a classical medium or even if it was composed of Newton's proposed "corpuscles", then, like the car example above, there ought to be a difference in measured speed from different directions. Nothing in their experiment squared with any known physical law, and the only thing they seemed to prove was that the speed of light remained constant regardless of the relative motions of the source and the observer. According to classical Newtonian physics, this should be impossible. In the end, it was Albert Einstein that bailed them out.

Relativity

While working at the patent office in Bern Switzerland between 1902 and 1905, Albert Einstein observed the serious discrepancies between Newton's laws and the discoveries of James Clerk Maxwell who had formulated a definitive, mathematically coherent theory of electricity and magnetism. Newton's laws could correctly describe the motion of material objects, but had proven to be worthless when it came to the behavior of electricity and magnetism, and of course, he knew from the Michaelson-Morley experiments that Newton's theories did not seem to apply to light either.

Maxwell had used the relatively new concept of fields to *unify* electric and magnetic forces, thus inventing the term "electromagnetism" to describe them. He discovered that self-propagating electromagnetic waves would travel through space at a constant speed, which happened to be equal to the previously measured speed of light, thus concluding that light was also a form of electromagnetic radiation. Maxwell's version of physics intrigued Einstein. Maxwell simply ignored the contradictions between Newton's version of physics and his own obviously successful theory. Maxwell's version of physics had entirely abandoned the Newtonian requirement for a compressible medium that allowed light waves to propagate.

At the time of his theory, no one realized just how groundbreaking Maxwell's insights were. According to classical physics, there was no daylight between our everyday reality and physical theory. Everything in Newton's theory "made sense". Newton's math may have been a mystery to ordinary people, but the reality that the math described could be understood by anyone because it conformed in every way to things that people observed every day. However, Maxwell's theory defied everyday logic. The math worked beautifully when used by engineers and scientists to describe what was happening on a microscopic level, but no one could actually describe the behavior of electrons and their associated magnetic field in ordinary English.

Maxwell had split the physical descriptions of the universe into two parts. One part described ordinary objects in Newtonian terms, and the other part described electromagnetic waves, including light, in mathematical terms. This was the key to modern physics. Even if the reality it describes makes no logical sense, mathematical descriptions of the universe could now be used by scientists to describe different physical realities provided that the realities they described conformed to experimental reality. And "realities" is the correct term. There are microscopic realities that in some unknown way fit together with our own to form our everyday world, but there are other, larger realities; not just microscopic realities, but universal realities as well.

Maxwell had released Einstein from the Newtonian paradigm that was retarding the advance of modern physics and allowed him to "think outside of the box". Einstein reasoned that physics should not have two entirely separate mathematical descriptions of reality. He spent the next three years trying to assemble a universal physical theory that could unify Maxwell and Newton and account for the motion of everything, including massive objects as well as light. The result was Einstein's *theory of special relativity*.

When Sir Isaac Newton formulated his theory in the seventeenth century, he assumed that mass, time and distance were constants. In other words, if an object weighed, say, one pound, and was two inches long, its weight and length never varies, no matter how fast it moves. This made sense and no one ever thought about any alternative. But now, with Maxwell's electromagnetic theory and the Michaelson-Morley experiment showing that there were two ways of explaining reality, everything was in a state of flux. All bets were off. No one had any way of reconciling the two realities--until Einstein!

Einstein was determined to tackle the discrepancies between the two realities, but in order to keep the speed of light constant, Einstein had to assume that mass, time and distance are not constants. The math simply did not work unless these quantities could vary depending on how fast the observer is moving relative to the object he is observing. It didn't seem to make sense at first, but the theory worked so well that he had no choice. He was mostly a theorist, and it was up to the experimentalists to prove him wrong. But even though the math described a reality that made absolutely no sense, the experimentalists did NOT prove him wrong! Much to the chagrin of many people, particularly the philosophers, Einstein's theory proved dead-on.

Although his theory is entirely mathematical, it has two great advantages:

- It is based on standard mathematics.

- Everything the theory has predicted has proven to be true in the real world.

Not all modern scientific theory has these two

advantages, and when we begin our review of modern cosmology you will see how important this is.

In 1915, ten years after formulating his theory of **special** relativity, Einstein advanced his theory of **general** relativity which integrated gravity and accelerated motion with the special relativity laws governing mass, time and distance. Einstein's concept of gravity involved the bending of space and time around massive objects. Understandably, none of this made any sense to the ordinary citizen of the day whose sense of reality didn't include traveling at speeds close to the speed of light, or the bending of space.

Relativity and Materialism

Einstein's laws of relativity were the first scientific crack in the philosophy of materialism. Now, neither time nor even matter were constants. This is why Einstein was able to write to a friend that *the distinction between past and present was only a "stubbornly persistent illusion"*. Now, the distance between objects could be different for different observers moving at different speeds, and even a feather might become as massive as a mountain if it is moving fast enough.

Since materialists base their beliefs on a completely rational understanding of matter, relativity began to throw their rock solid view of the cosmos into doubt. If time could be different for people traveling at different speeds, and mass could increase or decrease depending on the frame of reference, then the unchanging reality they had been relying on as the foundation of their beliefs was not as unchanging as they had thought.

Most materialists, however, brushed off these concerns because they seemed to apply mostly to the realm of the very big, and the very fast, and not to their own earthly frame of reference. In addition, Einstein's theory seemed so complicated that most people in the early twentieth century believed that no one could understand it but Einstein himself. (I'm old enough to remember how people in those days spoke about Einstein's theories.) Beyond that, the secular humanists took comfort in the fact that a lot of the things predicted by Einstein's two theories had not yet been verified, even by the time of Einstein's death in 1955. All you really needed was a few failures to show that his theories of relativity were wrong, and Newton was vindicated. Unfortunately (for the materialists at least), everything that Einstein's theories predicted later proved to be correct.

At the same time as Einstein was formulating his theories, another group of scientists from the northern countries of Europe were working on a second theory that would have devastating effects on the materialists' world view.

Quantum theory

Quantum theory is one of the two most successful physical theories ever proposed, and along with the theory of general relativity it has supplanted Newtonian classical physics. While Einstein's theories of relativity deal with the science of the very large and the very fast, quantum mechanics deals with the science of the very small. Einstein's ultimate dream was to create a theory that could unify all the laws of physics. Unfortunately, he was unable to unify his theory of general relativity with quantum physics, and that became a major sore point with him.

Einstein never totally accepted quantum theory, even though his early work was instrumental in creating it. He constantly argued with Niels Bohr, one of the fathers of quantum theory, because, in essence, Quantum theory describes a world in which reality isn't real.

Quantum theory grew out of Max Plank's formula for the power radiated by a "black body". A black body is any substance that absorbs radiation and then re-radiates it, (think of a red hot iron bar). *In order to create a formula which agreed with observed data, he came to the conclusion that energy comes in very small discreet packets.* The idea that light or heat actually varies by little jumps, rather than continuously, conflicts with our common sense view of everyday reality. Plank himself absolutely hated his solution, but no matter how hard he tried to find an alternative, he was never able to come up with one.

These discreet packets are called "quanta", and this new paradigm eventually led to the creation of a whole slew of laws which mathematically described the behavior of subatomic (elementary) particles, including the mysterious behavior of light, but made absolutely no sense when used to describe our everyday reality.

Once physics was released from the bonds of our normal analog (smoothly continuous) interpretation of reality, it could create mathematical laws that made sense when describing subatomic events. It also led to unexpected predictions of things that proved in subsequent experiments to be "real", but heretofore unimaginable. Without quantum mechanics we would never have invented our modern world of nuclear power, atom bombs digital products and lasers.

The diagram below is called a Schrodinger waveform. It's NOT a picture of how an electron or any other subatomic particle actually looks, but rather a representation of how it behaves. In essence, quantum mechanics paints a picture of the sub-atomic world in which particles exist only as "clouds of probability." Picture a cloud with a dense center which can look like a fuzzy ball or even a fuzzy dumbbell. Then picture the fuzziness grading off getting less and less dense around the edges as the distance from the center increases.

The cloud itself isn't the way, an electron, proton or any other sub-atomic particle really looks. It's just a probability field where the electron is *likely* to be found if a conscious observer actually tries to look at it. The electron is more likely to be found in the denser parts of the cloud and less likely to be found in the fuzzier parts.

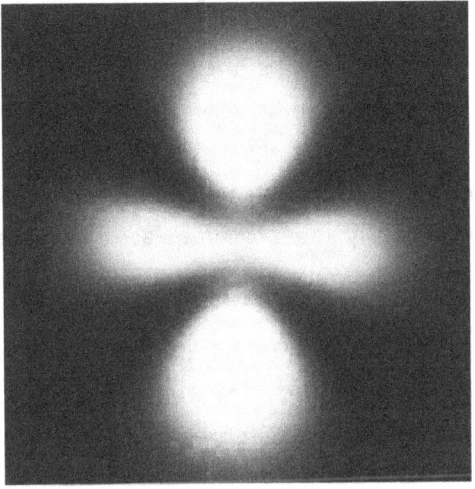

While Bohr, Heisenberg, Schrodinger and others were developing their theory of quantum mechanics, it became obvious that there was no way to create a mathematical device to transform a Schrodinger waveform into a discreet observable particle. Whenever anyone did an experiment that actually looked for a particle, they found one with measurable properties, but prior to actually observing it, they had to admit that the particle simply disappeared into a cloud of probability with no defined properties.

Mathematically, prior to the observation, the particle does not exist in any particular state. In fact, much to their chagrin, they had to admit that **until someone actually looks for them, discreet submicroscopic particles doesn't exist at all!**

Unobserved, all subatomic particles exist only as a statistical wave functions. The wave has an existence of its own and can be visualized if a scientist formulates an experiment looking for it, however, if the experimenter formulates an experiment looking for a particle, then the wave function disappears entirely and a discreet particle magically appears someplace within the area described by the wave function.

The mathematical formulas describing these clouds were worked out by Erwin Schrodinger in 1926. The image above is a graphic representation of an electron *probability* cloud as it exists around a hydrogen nucleus. It's not an image of an electron, but an image of where the actual electron is likely to be found if you or another *conscious* person go looking for it. Notice the term "conscious". It is, in fact, the key to understanding the true implications of quantum physics!

For hydrogen alone, there are 19 of these shapes depending on the excitation state of the electron. The shapes and the way they change are described by Schrodinger's equations.

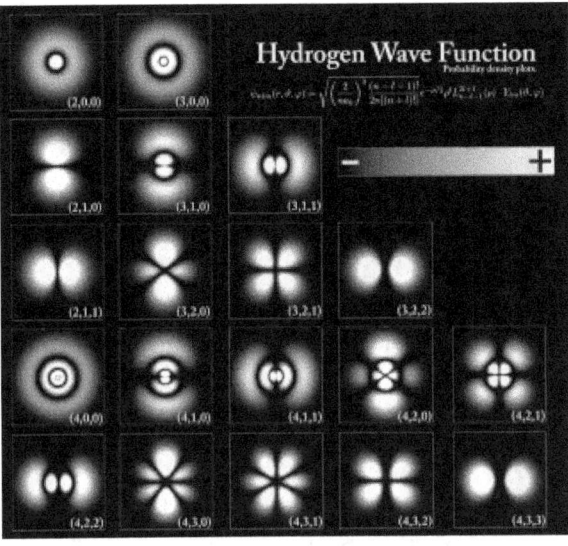

Keep in mind that neither Schrodinger nor his fellow scientists wanted to admit that human *consciousness* could become a part of an otherwise hard scientific theory. They all resisted it and at first assumed that it was just an artifact associated with their measuring instruments and methods of investigation. In the end, though, after a few more years of investigation by scientists outside of the original clique of Danes, Swedes and Austrians who invented quantum mechanics, everyone had to admit that *conscious observation* had to be included as a profoundly important part of the theory. Needless to say, modern physicists have fallen all over themselves trying to come up with alternatives, but they are all at least as outlandish as a consciousness based universe.

The Copenhagen Interpretation

The fact that there is no physical or mathematical way to create a physical particle out of its waveform without a conscious observer was considered to be a catastrophic development by the fathers of quantum theory. In an effort to evade this glaring defect, Neils Bohr and his fellow scientists developed what later became known as the *Copenhagen Interpretation* of quantum theory which places the collapse of the wave function within the observing instrument rather than in the brain of the observer. However, in 1932, mathematician John von Neumann destroyed this fig leaf in his groundbreaking book <u>The Mathematical Foundations of Quantum Mechanics</u>. We'll be discussing the Copenhagen interpretation in more depth later in this chapter.

There is a second, but rather outlandish interpretation of quantum theory that does not involve conscious observers. It is called "The Many Worlds Interpretation", and we'll be visiting it later.

2
Quantum Reality

Remember that everything in our reality is made from subatomic particles which have no real existence unless someone looks at them. This is NOT to say that matter isn't real. It's just that macroscopic objects like, say, a basketball, though real at all times, are made from tiny components which are not! This has very serious implications for the definition of our everyday "reality."

Everyone knows the old solipsism "If a tree falls in the forest and no one is around to hear it, does it make a sound?" Quantum physics now forces us to ask if the tree even exists at all if no one is around to see it. While this may seem like a frivolous philosophical speculation, it has enormous implications. Every branch of the hard sciences is based on the fundamentals of physics. As an example, biology is based on biochemistry which is based on organic chemistry which is based on inorganic chemistry which is based on physics involving the interactions of atoms, which are themselves composed of quantum particles, and which aren't really there unless you look at them. None of this makes any sense, of course, but it is an acknowledgement that our everyday reality stands on much more shaky ground than we generally realize.

Science, Math and God

This idea has rather ancient roots. Bishop George Berkeley, who died in 1753 was the originator of a philosophy known as **immaterialism**. This theory denies the existence of material substance and instead contends that familiar objects like tables and chairs are only ideas in the minds of perceivers, and as a result cannot exist without being perceived. Until the twentieth century (and the advent of quantum physics) he was widely ridiculed, and in the quadrangle of Balliol College, one of the constituent colleges of Oxford University in England stands a plaque with this rather famous limerick:

There was a young man who said, God
Must think it exceedingly odd
If he finds that this tree
Continues to be
When there's no one about in the Quad

And beneath it, there is this response, by the Catholic theologian and Bible translator Monsignor Ronald Knox:

Dear Sir, your astonishment's odd:
I am always about in the Quad.
And that's why the tree
Will continue to be
Since observed by, Yours faithfully, GOD.

How to avoid a cosmic consciousness without really trying

The point in all of this is that quantum physics is the most comprehensive and accurate description of nature invented by the mind of man. It implies that *consciousness is an intrinsic property of our universe*, and not just a byproduct of the human brain. This is probably the most profound implication of quantum physics, yet many influential scientists refuse to acknowledge it. (This is primarily because of the overriding influence of Secular Humanism within Western institutions such as academia and the media. Secular Humanism is discussed in chapter 9.)

In the modern era, there has been a mad scramble to find a theory...any theory...that can explain the universe without having to resort to a cosmic observer. The paranoia has reached hilarious proportions. In 2013, an article in New Scientist Magazine highlighted an idea that was so bizarre that it is hard to believe that it was ever seriously considered:

"LEGIONS of disembodied brains floating in deep space threaten to undermine our understanding of the universe. New mathematical modelling suggests string theory and its multiple universes may just provide our salvation – and that could win the controversial theory a few more backers.

Science, Math and God

Physicists have dreamed up some bizarre ideas over the years, but a decade or so ago they outdid themselves with the concept of Boltzmann brains – fully formed, conscious entities that form spontaneously in outer space.

It may seem impossible for a brain to blink into existence, but the laws of physics don't rule it out. All it requires is a vast amount of time and space. In other words, an infinity of time and space. Given an infinite amount of time and an infinite amount of space, eventually, a random chunk of matter and energy will happen to come together in the form of a working mind. It's the same logic that says a million monkeys working on a million typewriters will eventually replicate the complete works of Shakespeare if you leave them at their typewriters long enough.

Most models of the future predict that the universe will expand exponentially forever. That will eventually spawn inconceivable numbers of Boltzmann brains, far outnumbering every human who has ever, or will ever, live."

According to most cosmological theories, the age of the universe is approximately 13.8 billion years (although the most recent research places it at 26.7 billion years). This means that any astronomical bodies that we can see with our present technology must lie within a radius that allows enough time for their light to reach us. The furthest galaxy we have seen so far is HD1 which lies 13.5 billion light-years away from us. However, all galaxies are receding away from us at speeds even exceeding the speed of light, so in the length of time that it took the light from HD1 to reach us, the actual distance from us may now be more than twice that distance. Even so, that does NOT mean that the edges of the universe are a "mere" 27 billion light years away from us. The fact that we cannot see anything beyond that distance doesn't necessarily mean that that is the full extent of the universe. We cannot see objects further than 13.8 billion miles away because light simply hasn't had enough time to reach us from further away. 13.8 billion miles is simply the radius of the *observable* universe. We have no idea of exactly how big the wider universe actually is. (Note: The galaxies may be receding from us faster than the speed of light because their acceleration is caused by the expansion of the SPACE between them, and not be any explosive momentum caused by the big bang itself.)

Data from several satellites (PLANCK and WMAP) have shown that the universe is almost certainly FLAT. This means that it isn't curved around like a sphere with hard edges separating the universe from whatever lies beyond. For all we know, the universe may go on forever, and considering the difficulties that the big bang theory has encountered in recent years, it may well have been in existence infinitely longer than 13.8

billion years, maybe even forever. In other words, both time and space may be infinite.

It is interesting to note that infinity has become the gift that keeps on giving in modern cosmology. Given an infinity of time and space, absolutely NOTHING is impossible, including the spontaneous appearance of disembodied conscious brains. **However, given all these infinities, why the creation of an army of Boltzman brains should be a certainty, but the spontaneous appearance of an all-knowing god remains impossible (according to modern secularists) remains a bit of a mystery**.

The "new mathematical modelling" mentioned in the first paragraph of the article above remains just that...mathematical modelling. (See chapter 4 to review superstring and M theory.) As regards the concept of Boltzmann brains, one has to marvel at the ingenuity of modern scientists to discover ways of justifying atheism.

But, as we've seen, resistance to the idea of a "Cosmic Observer' did not begin in the twenty first century. It's been around since Bohr and his grad student Heisenberg invented the concept of the Copenhagen interpretation of quantum reality. Copenhagen allowed scientists to ignore the implications of quantum physics, and about the only scientist that wanted to discuss these implications was Einstein whose arguments with Bohr are legendary.

Quantum Entanglement

The property that later became known as "quantum entanglement" was first addressed by Albert Einstein and his collaborators, Boris Podolsky and Nathan Rosen in 1935. Einstein and his collaborators wrote a paper in which they presented a thought experiment later called the *EPR Paradox*.

At the time, Einstein was arguing with Niels Bohr about the validity of the newly minted theory of quantum mechanics. Einstein never wanted to believe that the objects described by quantum mechanics did not exist unless they were observed. However, he discovered something else that quantum theory predicted which struck him as even more impossible.

Einstein and his colleagues were the first scientists to note that the raw mathematics of quantum theory predicts that two sub-atomic particles which physically interact, or were created in the same event are linked in such a way that, after their interaction, any measurement made to discover a quantum property in one particle will *instantaneously* produce the correlated value in its mate, no matter how far apart they are. This type of communication is termed "superluminal" because it happens "instantly" across any distance, faster than the speed of light. This instant linkage between two separated particles is called *quantum entanglement*.

According to Einstein's well-worn theory of relativity, nothing can travel faster than the speed of light. This instantaneous communication between entangled particles made absolutely no sense, and Einstein would have none of it! "Now I have him!" thought Einstein. "Bohr's beloved quantum theory predicts something impossible, so the whole theory must be wrong."

Einstein's own theory of relativity clearly states that instantaneous communication between two particles is physically impossible, and he jokingly called this instant communication "spooky action at a distance." He was not happy with any of the implications of quantum physics and was hoping to find a fatal flaw in the theory. He now believed that he had finally found one.

It turned out that Einstein was (finally) proven wrong. It was later shown that quantum entanglement does, in fact, exist. Quantum entanglement is now so well integrated into quantum theory that it is even being used in a new generation of "quantum computers."

As an example, take the quantum property known as "spin." Quantum properties like spin are not like their real world counterparts. In the real world, the spin of a top might be clockwise or counterclockwise depending on whether you view it from above or from below. A quantum particle, on the other hand, will always have the same spin no matter how you look at it. Quantum spins can be either up or down.

Quantum theory says that no particle can have a spin or any other property until it is observed. This, of course is obvious because the particle doesn't even exist unless someone observes it. Once observed, the spin has an even chance (50/50) of being in either state. This *"quantum randomness"* is another absolute law of quantum mechanics. No one can *predict* which spin any given particle will have until it is actually observed. This cosmic randomness was just one more piece of the theory of quantum mechanics that Einstein hated. He once wrote to a friend "The [quantum] theory says a lot, but does not really bring us any closer to the secret of the 'Old One'. I, at any rate, am convinced that He does not throw dice." (Yes, Einstein believed in God!)

Technically, this quantum randomness should apply to both members of an entangled pair. Since neither particle exists as a real entity until it is observed by a conscious being, then, in our normal material world, measuring the spin of one particle *should* have no effect on the measured spin of its unobserved mate. However, experiments show that this is not the case.

Quantum theory predicts that any two entangled particles will always be found to have complementary spins no matter how far away they are from each other, even when the experimenter performs a blind reversal on the first particle. That means that if you observe the spin of one member of an entangled pair, you CAN predict the spin of its mate with 100% accuracy. This is a clear violation of the 50/50 quantum law stated above.

Quantum theory clearly states that communication between the particles takes place instantaneously, just as though there is no distance between them, even if the particles are on opposite sides of the universe. In other words, entangled particles behave as though they are not separated in space, just as as if they were the same object, no matter how far apart they are. But in our own reality, these objects are objectively not even near each other, so *how can two separate particles, not in direct contact, transfer information instantaneously*? Einstein's theory of relativity clearly states that nothing can travel faster than the speed of light, so how could this be possible? Clearly, either Einstein or quantum theory must be wrong!

Bell's Theorem

Because Einstein believed in the impossibility of superluminal transfer of information, the EPR Paradox paper argued that although quantum mechanics was correct as far as it went, it was incomplete. Einstein thought that the weirdness of the reality described by quantum theory was actually due to "influences" that were hidden in the mathematics, but actually existed in the real world. He felt that in time, these *hidden variables* would be elucidated and prove that electrons, protons, photons and all the other members of the sub microscopic menagerie are real particles with or without conscious observation.

Einstein once jokingly asked a colleague if he believed that the moon was only there when he looked at it. He simply couldn't believe in a quantum theory that implied that our classical reality was founded upon a deeper reality in which a material particle does not exist unless it is observed by a conscious entity. He once wrote to a colleague: "The Lord God is subtle, but malicious He is not."

In 1964, physicist John Stewart Bell looked at all the experimental evidence pertaining to the theory of quantum mechanics and proposed a new theorem. This theorem set up conditions which *had to be true* of any world in which, like our own, the reality of objects and their separability in space (locality) was true. In addition, these conditions were experimentally testable. Bell had discovered a way of proving or disproving that matter, as we know it, can exist without a conscious entity observing it. In other words, he defied physicists to prove that our reality is "real"!

Bell found that *if our* reality extends all the way down to the level of quantum objects, then certain observable quantities had to be larger than others. This disparity in the size of quantities is called *Bell's inequality*. If the inequalities were found to exist, then our material reality is based upon a submicroscopic world in which objects exist all the time, and are separated in space regardless of whether they are observed or not. In other words:

> If Bell's inequality is true (i.e. if these inequalities exist and these observable quantities ARE larger than others) then quantum physics is wrong or at least incomplete, and quantum particles not only exist as "real" objects (like the billiard balls in the above example), but are also separated in space (like our billiard balls) regardless of whether they are observed or not. In other words, our everyday reality extends to the world of quantum particles, quantum theory is wrong, conscious observation is not necessary to create our reality, and Einstein was correct.

> If Bell's inequality is false, (i.e. if the observable quantities are NOT larger than others) then quantum theory is correct in every aspect, and electrons, photons, quarks and all the other submicroscopic entities never come into reality as actual particles unless a conscious entity observes them and Einstein was wrong..

> Furthermore, if Bell's inequality is false, then quantum entangled particles objects, are NOT separated in space! In other words,

space itself is a fiction! This property of the universe is called non-locality and is covered under a separate heading later in this chapter.

In 1972, John Clauser and Stuart Freedman, (and again in 1981, Alain Aspect) performed the requisite experiments and discovered that Bell's inequality IS indeed violated (false), consciousness IS an intrinsic property of our reality and Quantum physics *is* correct in every aspect. In other words, our everyday reality stops upon entering the quantum realm. In more succinct language:

> *There is no way that we can ever correlate the reality implied by quantum theory with our everyday reality. Or, in more down-to-earth language, "Don't believe your lying eyes!"*

This means that **quantum objects do not exist unless they are observed by a conscious entity**. It also means that our concept of locality, which is the separation of objects in space, is also an illusion. Material objects are, of course, made of quantum objects, and therefore must also require some form of conscious input in order to exist. It turns out that we really live in two separate realities at the same time, and those realities are not only separate, but incompatible!

All of this is rather unsettling. We can hypothesize that the obvious persistence of the material objects that we perceive in our everyday lives, (as well as their separation in space) *may* have something to do with the enormous number of interacting quantum objects of which they are composed. This "bridge" between an observable object (say, a basketball) and the quantum particles that compose it would, in all likelihood, involve quantum entanglement between the multitudes of particles that compose it. Unfortunately, there is no hint within the mathematics that quantum theory predicts this effect. Therefore, the mystery of how a bevy of "unreal" waveform/particles can somehow create a real world object persists.

No one really knows how we can get from the "unreal" world of quantum objects to the persistent reality we live in. Nor does anyone understand just why *conscious* observation is essential to the creation of matter, but in proving that quantum reality is the ultimate reality, **Bell proved that without conscious observers, nothing can exist as a physical entity.** Bell's theorem and the experiments that followed it proved that all matter, from photons to mountains, are ultimately created by conscious observation.

Perhaps Bishop Berkeley was correct!

Note: we will be discussing an alternative explanation for the consciousness/reality nexus later when we compare the Copenhagen Interpretation of quantum physics with the

Many Worlds Interpretation.

Heisenberg's uncertainty principle

Werner Heisenberg was Neils Bohr's grad student. His doctoral thesis was based on a thought experiment called the Heisenberg microscope. In the end it evolved into a mathematical proof demonstrating that the determinate universe predicted by *classical* physics is "impossible".

Momentum and position are two of several different complimentary properties that can define a quantum particle. Heisenberg proved that the measurement of any quantum property would invariably perturb (change) its complimentary mate making it impossible to know both properties at the same time. Therefore, if you measure the position of an electron, you can't measure its velocity at the same time because (Heisenberg reasoned) the measurement of the position would have given the electron a "kick", thus changing its direction and speed. If you measure its velocity, then that measurement would have changed its position. Heisenberg's uncertainty is more accurately phrased thus: The more accurately you measure an objects position, the more uncertain you will be about its velocity; the more accurately you measure an object's velocity, the more uncertain you will be about its position.

In the years since Heisenberg made his discovery, it has become clear that the uncertainty principle is not associated with any energy transfer from an observation. It is, in actuality, inherent in the properties of all wave-like systems. That is to say that Heisenberg's uncertainty principle is another intrinsic element of the material universe, right beside the essential nature of consciousness in "producing" matter.

Remember that in Newton's classical clockwork universe, if we could know the mass, position, direction and speed of all the particles in the universe, then we could predict the future with absolute accuracy, and human free will would be a myth. This axiom became a truncheon that materialists used for 240 years to undermine religion's assertion that God gave man the gift of *free will* (from the publication of Newton's Principia in 1687 to Heisenberg's discovery of the uncertainty principle in 1927). If everything was pre-determined by the mass, position and motion of the atoms in the universe, then free will was impossible.

After the publication of Heisenberg's uncertainty principle, the materialist position became entirely untenable. Unfortunately, the materialists never noticed.

To read more about the Heisenberg uncertainty principle, visit the website of The American Institute of

Physics. Avoid Wikipedia on any subject that pertains to spirituality. Virtually every relevant page has been edited by CSICOP (Now renamed as CSI) which is a vigilante Secular Humanist organization dedicated to denigrating or entirely eliminating any evidence that does not promote materialist philosophy. The last chapter of this book is dedicated to CSICOP and debunking the claims of modern skeptics.

Non-Locality

The concept of locality, or the separation of objects in space is an interesting, but slippery subject. The Bell's Theorem experiments proved that instantaneous action or transfer of information between entangled particles does indeed happen. A recent milestone relating to research in this area took place at Delft University in 2015. According to the NY Times:

> "The new experiment, conducted by a group led by Ronald Hanson, a physicist at the Dutch university's Kavli Institute of Nanoscience, and joined by scientists from Spain and England, is the strongest evidence yet to support the most fundamental claims of the theory of quantum mechanics about the existence of an odd world formed by a fabric of subatomic particles, where matter does not take form until it is observed and time runs backward as well as forward.

The researchers describe their experiment as a "loophole-free Bell test" in a reference to an experiment proposed in 1964 by the physicist John Stewart Bell as a way of proving that "spooky action at a distance" is real."

Aside from proving that Einstein was wrong in his initial assessment of quantum mechanics, experiments like this also shake up our everyday, materialistic understanding of *locality*. Superluminal transfer of information implies that in spite of the fact that we perceive objects (and people) to be miles apart, the distance between them from a quantum point of view is actually zero! This is a concept called *"quantum non-locality"* which is just another way of saying that distance has a rather nebulous meaning in the world of quantum physics.

Ultimately, our normal understanding of "distance" is actually an illusion. If the distance between entangled particles at opposite ends of the universe is shown to be zero, then it follows that the distance between ANY two particles is also zero! After all, Physicists believe that the universe started out as a singularity, and almost all subatomic particles were formed in the same event when the density of matter was extremely high. Therefore, virtually all quantum entities formed during the "Big Bang" when the cosmic singularity exploded. The Big Bang is assumed to be the event that began the universe and will be explored more fully later.

In another example of non-locality, imagine that a photon is created on a star located light years away from the earth. Remember that a photon in its virgin state is not a particle, but rather a wave, and its waveform would spread out at the speed of light over the time it takes to reach us in a wide, ever-expanding sphere with a surface area hundreds of billions or even trillions of light years wide...until one day it encounters the retina of your eye. In that instant, the entire spherical wave collapses into a single sub-microscopic particle (a photon). Once you observe it, it vanishes from the universe, and no longer exists, even though just a moment before, it existed throughout billions of miles of outer space. The collapse of the waveform does not happen at the speed of light, but rather instantly. In other words, it is superluminal and defies Einstein's laws of relativity. This is another instance of non-locality in action. Keep in mind that your eye could have been located on a moon of Jupiter, and the same photon could have instantly materialized into a photon there.

Quantum non-locality is now a tenured part of quantum physics (although it is still hotly debated with most experimental physicists simply accepting its reality and incorporating it into their calculations while some theoretical physicists reject it because it does not fit into any common sense view of the universe. The arguments at quantum scientific conferences get very heated, but no one ever wins!) In any event, quantum physics underlies the reality we live in, and therefore, quantum non-locality implies that *nothing in the universe has a definite location, or alternately that everything is in contact with everything else* (which is exactly why many theoretical physicists reject it.)

In his easy-to-read book <u>Spooky Action at a Distance</u>, George Musser gives us just a glimpse of its reach:

> *We've seen non-locality pop up all over the place: in experiments on the quantum realm, in the paradoxes of black holes, in the grand structure of the universe, in the maelstrom of particle collisions. In all these examples, physics enters a twilight zone. Things can outrun light; cause and effect can be reversed; distance can lose meaning; two objects may actually be one. The universe becomes spooky.*

Non-locality can be taken to mean that every submicroscopic particle, and therefore every macroscopic object composed of submicroscopic particles actually fills the entire universe. If every sub-atomic particle is in contact with every other sub-atomic particle, then every motion, every behavior, every human thought could have an effect on absolutely everything else instantaneously, no matter how far away it may appear to us.

Reality, as we commonly perceive it, is actually an

illusion! In spite of the illusion produced by our senses that the universe is billions of light years across, from a quantum point of view, it really is much smaller than the period at the end of this sentence. Even though you live in LA and Aunt Martha lives in Boston, you both may occupy the same space. Talk about "spooky action at a distance! This is not some sort of mathematical fluke that has no real meaning in the context of our everyday lives. Quantum mechanics, in all its esoteric strangeness, is an exact mathematical description of the very foundation of our material existence! It is the super-reality that underlies the reality we live in every day! (Keep this principle in mind later when we study ESP in chapter 7.)

Consciousness and the Copenhagen Interpretation,

As we saw in chapter 1, the most basic tenant of quantum mechanics is that the state of an elementary particle is completely indeterminate until it is actually observed by a conscious observer. Instead, it exists only as a mathematical probability waveform. In other words, all of its properties like location, velocity and spin exist *simultaneously* in a cloudlike state. When opposing properties exist simultaneously, it is said that they exist in a *superposition* state. By "opposing properties", I mean things like up or down, positive or negative, here or there and even living or dead. Opposite competing properties cannot exist at the same

43

time in our normal everyday reality, but in the world of quantum mechanics they can.

The wave-particle duality is the reason for this state of affairs. Prior to being observed, a subatomic particle doesn't actually exist as a particle, but instead it exists as a waveform with the *potentiality* to become a real particle. All of the particle's properties exist in a superposition state within that waveform. Only when someone actually tries to observe the quantum particle does its *wave function collapse.* At that instant, all of the properties originally combined in the superposition state "magically" separate and the particle pops into reality. According to the originators of quantum theory, the particle only gains properties like mass, velocity, location, spin, polarization...etc. after observation by a conscious entity and the subsequent collapse of its waveform.

The Copenhagen Interpretation contains an implicit philosophical argument that emerged directly from quantum observations. The world it described was so fantastically bizarre that an overall description became necessary to allow the public (and budding scientists) to cope with the discovery that our everyday reality is underlain by an irrational and totally inexplicable reality at the most basic level. An engineer or physicist may be content to work strictly with the mathematics of quantum mechanics, but most ordinary people want to understand what it means!

The Schrodinger equation predicts the smooth evolution of a wave function over time. It predicts the shape of the wave as it propagates through infinite space, but does not predict the collapse of the wave into a discreet particle upon observation by a conscious observer. Furthermore it does not address the question of how an infinitely large wave can collapse *instantly* into a discreet particle. The math simply stops working whenever someone observes a quantum system with a scientific instrument. In order to mathematically describe the sudden collapse of the wave function whenever it is observed, physicists are forced to add an ungainly mathematical postulate which is not derived from the original theory.

The Copenhagen interpretation was formulated by Niels Bohr and his grad student, Werner Heisenberg between 1925 and 1927. It includes the instant collapse of an infinitely large wave into a discreet particle whenever a conscious entity tries to measure any of its properties such as its location or momentum. This concept, undeniable, based on experimental observation, horrified both Einstein and Bohr both of whom wanted a way to nullify the need for conscious observation demanded by the theory. Bohr's solution was simply to ignore consciousness entirely.

Einstein's approach was to dissect non-locality in an attempt to debunk the entire theory. According to Einstein's theory, superluminal travel is impossible, and the nullification of space is just plain nuts!

However, whether ignored or dissected, until someone could disprove these unsettling tenants of quantum theory, both Bohr and Einstein were stuck with the horrifying fact that in the world of the very small:

> 1.   *No physical particle actually exists unless it is observed by a conscious entity.* The only thing that exists as a reality on the unobserved quantum level is a *wave function* which defines the statistical probability of *finding* a particle in a particular location.

> 2.   Space either doesn't really exist or it has a thoroughly different meaning than our everyday perception of it.

> 3.   Time as we understand it does not exist on the quantum level. It may even run backwards.

Bohr had no interest in trying to explain quantum reality to an uneducated public. When forced to make a statement on the subject, he had this to say:

> *"There is no quantum world. There is only an abstract quantum description. It is wrong to think that the task of physics is to find out <u>how</u> nature is. Physics [is only concerned with] what we can say <u>about</u> nature."*

Note that this view is a stark deviation from the definition of science that had been in effect since Aristotle first invented the concept. Until 1925, the goal of science had always been to *describe* nature and make its inner workings understandable. After Bohr, however, the ultimate description of nature became a massive mathematical formula. From that point on, the inner workings of the physics became completely incomprehensible except in terms of the math.

Bohr's solution to quantum physics' "reality problem" was, in the end, an evasion of the implications of quantum mechanics. It was (and still is) meant to reconcile the microscopic world which is governed by the statistical laws of quantum mechanics, with the macroscopic world, which is governed by Newton's classical laws of physics, without offending the bulk of other scientists. This modern implication that reality required consciousness to exist was a huge step away from the scientific progress made, since the reformation, toward removing man from the center of the universe!

Bohr and his crew knew that he was defying the rationalists in charge of the scientific establishment with his new theory. Therefore, in order to avoid censure and assuage his own misgivings, he created an escape hatch. He noted that we cannot *directly* observe any object in the quantum world. In order to observe them, we must resort to macroscopic instruments, and these instruments exist in the real world described by Newton's classical physics.

In other words, the Copenhagen interpretation originally assumed that the instruments are doing the observing, and not us. After all, only the instruments can make the actual measurements. We, as conscious observers are only reporting what the instruments are telling us. In other words, a conscious observer is not collapsing the quantum waveform. A mechanical instrument is collapsing it. It isn't an ideal explanation, however it turned out to be a great excuse for physicists to go about their work without ever having to consider the implications of quantum theory. In Bohr's own words, "It was an act of desperation!"

Remember that Einstein began his quest for the theory of general relativity because he believed that science should not have two separate theories of physics. Bohr had found a way to rationalize doing the exact opposite with reality. Using the Copenhagen interpretation, entire generations of physicists have been taught that the equations of quantum mechanics work only in one part of reality, the world of the microscope and mathematics, while ceasing to be relevant in another, the macroscopic.

However In his 1932 book, The Mathematical Foundations of Quantum Mechanics, John von Neumann came to a very different conclusion. He presented a rigorous mathematical treatment that proved that while the Copenhagen interpretation was absolutely correct in all its predictions, the assumption that it described a world entirely separate from our own is wrong. His mathematical treatment proved that quantum mechanics applies to both the quantum and classical worlds, and the collapse of the wave function happens at the point along the chain of observation when it comes into contact with consciousness, not when it comes into contact with a "dead" measuring instrument.

Since von Neumann's time, his mathematical description of the quantum world has become the accepted standard. Eugene Wigner and others have shown, using Von Neumann's math, that the consciousness of an observer is the demarcation line which precipitates collapse of the wave function. From this point of view, the measuring instruments themselves enter a superposition state along with the quantum particle they are measuring, and the superposition state only collapses when the instruments are finally read by a conscious observer. These scientists have undermined Bohr's original assumption that placed the demarcation line between the quantum particle and the measuring instrument.

The enormous success of the quantum theory has proven beyond doubt that the experimenter must now be thought of as a fundamental part of the experiment rather than an impartial and distant referee. The observer's decision to formulate an experiment in the way he or she does determines the type of outcome that will be observed. If there is no one to observe, there is no outcome! Our classical world is built from quantum objects. Although scientists don't want to admit it, Von Neumann's math has enormous implications. It's very possible that **the existence of our reality depends upon the existence of conscious entities that can observe it!**

This may seem to defy logic, and indeed, there are non-mathematical arguments against it, but never before 1925 had anyone in the world been able to make that statement with any evidence to back it up. Today, we can! Maybe Bishop Berkeley was right. If a tree falls in the forest without a conscious observer, not only does it NOT make a sound, but maybe it doesn't even exist! Well.....Maybe!

Schrodinger's Cat

Even Erwin Schrodinger, like most scientists had real problems with the Copenhagen interpretation, even though he had been instrumental in formulating it. It was he who created the Schrodinger equations which are still used to describe the evolution of quantum waveforms. In 1935, he set up a "thought experiment" placing a sort of "Rube Goldberg" apparatus in a large box, along with a cat. The apparatus is composed of a device with a radioactive element linked to a hammer that may or may not slam down breaking a glass vial full of poison gas. If the hammer strikes the glass vial, it releases the gas killing the cat. But here's the hitch! Whether or not the hammer hits the glass vial releasing the gas depends on whether or not the radioactive element randomly decays.

Since the state of the radioactive element remains indeterminate until it is observed, the Copenhagen Interpretation of quantum theory says that the state of the hammer, the vial of poison gas, and the cat are also indeterminate until someone looks at them. The cat, therefore, exists as both dead AND alive until someone peaks into the box to find out which state the cat is in. In more technical terms, the radioactive element, the hammer, the vial of gas and the cat are all in a *superposition state.* According to quantum physics, all of them remain in this in-between state until they are observed, after which, all wave functions collapse into a definite state; i.e. the cat is either dead or alive!

The object of this exercise was to show that quantum theory was either incomplete, or dead wrong! Schrodinger intended his paper to enlarge on Einstein's EPR paper. "Ha Ha, see? A cat can't be both dead and alive at the same time, and if the cat is in one state or the other, then the quantum state of the radioactive particle which is at the root of the experiment must also have a definite state regardless of whether anyone looks at it or not!"

Unfortunately for Schrodinger, he was wrong. In 1965, Richard Feynman created a thought experiment to show that individual quantum particles could be in two complimentary states at once, and would remain indeterminate until observed by a conscious observer. His experiment would probe the reality of the wave/particle duality of quantum particles, which until that time was understood only as theory.

Ten years later, in 1975, Feynman's experiment was successfully carried out. Quantum theory was shown to be correct, and Schrodinger's famous suspicions were finally put to rest. Apparently, cats CAN be both dead and alive at the same time (or something like that)!

The story of Schrodinger's cat has implications well beyond the life or death of the cat itself. Finding the cat dead would **create a *history*** of its developing rigor mortis; finding it alive would **create a history** of its developing hunger—*backward in time*. The fundamental laws of quantum physics are completely neutral with regard to the direction of time. Quantum mechanics predicts that a conscious observation of any event or object can actually create a *history* that is consistent with the observation. Note that the observation must be by a CONSCIOUS entity, and that time itself can run backwards.

The implications of quantum theory troubled Einstein greatly. In his book <u>Quantum Enigma</u>, Bruce Rosenblum tells of his meeting in the 1950's with a fellow physics graduate student at the home of Albert Einstein.

> *"Soon Einstein asked about our quantum mechanics course. He was pleased that we used David Bohm's text and asked how we liked Bohm's initial philosophical treatment. We couldn't answer. We'd been told to skip that part of the book and concentrate on the section titled, "The Mathematical Formulation of the Theory." Einstein persisted, but the issues that concerned him were unfamiliar to us. Our training was in the use of the theory, not its meaning. Our responses disappointed him, and that part of the conversation soon ended. It would be many years before I understood Einstein's profound concern with the mysterious implications of quantum theory, implications that he called "spooky" and that he believed denied the obvious existence of the real world."*

Kuttner, Fred; Rosenblum, Bruce (2008-10-14). <u>Quantum Enigma</u>: Physics Encounters Consciousness (Kindle Locations 178-183). Oxford University Press. Kindle Edition.

Even today, it is virtually "politically incorrect" for physicists in academia to discuss the implications of quantum theory. Most physicists find these implications too "theological" to entertain, so they teach the USE of quantum mechanics only. But whether these implications are suppressed or not, they certainly exist and they support our perception of free will in no uncertain terms. They demand that consciousness be placed at the center of any theory of the universe!

In any case, quantum theory proves that *REALITY is not as simple or definite as we assume it to be.* What we perceive to be reality is underlain by a reality much deeper than the one we see. Note that it was Schrodinger who coined the term "quantum entanglement." He was one of the fathers of quantum theory, and yet, even he didn't want to believe in its implications!

The two competing theories of quantum reality

The rather confusing picture of quantum reality drawn in the first two chapters has been interpreted in two different ways. These interpretations became necessary, first because it is virtually impossible for an ordinary person using everyday logic to understand the underlying reality revealed by observations of the quantum world, and second, because no one has yet been able to reconcile quantum theory with relativity theory. The unification of these theories into a single

Grand Unified Theory has become a major goal in the world of physics and cosmology.

The first and oldest interpretation of quantum theory was, surprisingly, to have no interpretation at all. **The Copenhagen Interpretation**, which we have touched upon previously, was formulated by Niels Bohr and his Danish colleagues and is the mathematical foundation of the entire quantum theory. The fact that the mathematics not only explained all known observed quantum phenomena but also predicted many other unexpected phenomena made the theory one of the three most important discoveries in the history of physics (classical physics and relativity being the other two). The fact that quantum phenomena themselves made absolutely no sense in the light of everyday reality caused not a ripple of worry among the fathers of quantum mechanics.

Maxwell had simply ignored the tenants of classical physics when he discovered the principles of electromagnetism. Einstein had followed in his footsteps and ignored our ordinary understanding of mass, time and distance. Therefore, it wasn't much of a leap for Bohr and his colleagues to simply ignore the rest of reality when formulating their theory. Even the fact that consciousness had reentered the cold world of physics was simply something that could be placed on a back shelf and ignored. The theory and its functional uses were what counted. The Copenhagen Interpretation explicitly ignores the centrality of consciousness in quantum theory in the same way that Maxwell, Einstein and Heisenberg ignored the dictates of Classical Physics when formulating their respective theories.

Still, the ramifications of quantum theory caused tremendous anxiety in the minds of many modern academic scientists because, even though it was not understood by the public at large, consciousness had wormed its way back into the center of the universe where it had resided for the countless millennia from the time God said "Let There Be Light" until Sir Isaac Newton said "Light is made of corpuscles". Newton had banished God and even man from the center of the Universe. Now, all of a sudden, consciousness, in all its ineffable subtleness seemed to have shoved itself unceremoniously back where it was not wanted. This contravened the secular drift in academia that existed since the publication of Newton's Principia.

Aristotle's earth centered universe placed mankind at the epicenter of the universe because humans were assumed to be the gods' crowning achievements, and religious organizations since that time have always echoed this sentiment. However, after Newton's classical physics took root and the Catholic Church's embarrassment after "the Galileo affair", the academic establishment took every opportunity to deny mankind's privileged position. Quantum theory was a severe blow to this secular trend, however they were more than willing to embrace quantum physics so long as no one mentioned consciousness.

But without a conscious observer, the wave never collapses into a particle, and the centrality of consciousness in quantum theory always lurks in the back of every physics student's mind. Partly for this reason, we now have a series of alternatives to textbook quantum theory and its Copenhagen Interpretation. They are collectively called **"The Many Worlds" interpretations** and they will be discussed in detail in chapter 4.

The Many Worlds interpretations are NOT based upon experimental observations and rely exclusively upon alternate forms of mathematics which, if properly tweaked, can be forced to describe the same quantum world described naturally by standard quantum theory. Unfortunately, they also make cosmological predictions that can never be confirmed through earthly observation. The reason that no observations can be made is because most of the predicted phenomena either lie in dimensions beyond our material existence or predict quantum entities that are too infinitesimally small ever to be observed in any real-world particle accelerator.

Two of the new mathematical inventions are **string theory** and its successor **M theory**. These theories have gained their popularity with academic scientists to a large extent because they have the *potential* to unify gravity with the other four forces of nature. They also predict a series of interesting alternatives to our single universe called **multiverses**. An infinite number of universes are really quite convenient for cosmologists because they can be used to explain why 25 arbitrarily set universal constants are so finely tuned that they allow life to exist at all. (The universal physical constants and the multiverses that spawn them will be discussed in the next chapter.) One of the earliest multiverse theories even eliminates the need to place consciousness and hence life into a position of importance in an otherwise cold and chaotic universe.

3
Cosmology; The new creationism

Ancient theories of the universe

The ancient Greeks believed in an Earth centered universe with the stars, the moon and the sun moving around the earth on a series of revolving concentric "perfect crystal (transparent) spheres". All celestial objects were affixed to these spheres except the planets which often moved in "retrograde" motions, meaning in reverse of the direction that the sun, the moon and the stars moved. The planets were assumed to be "wandering" objects only loosely attached to their crystal spheres and moved by the gods. This system worked well for the mindset of the times because anyone could look up and see the way the stars, the sun, the moon and the planets moved in the sky. It could be argued that the these perfect heavenly spheres and the planets were created and manipulated by the gods who lived in the heavens above, while we lived on an earth that was at the center of the heavens. Thus mankind, the ultimate achievement of the gods, was the center of all creation.

In about 100 A.D., a Roman astronomer named Claudius Ptolemy showed that the motions of the planets could be accurately predicted (mathematically) if he assumed that they were attached to smaller crystal spheres that were themselves attached to separate earth centered crystal spheres. These smaller spheres were called "**epicycles**". The invention of a mathematically predictable universe not only allowed a newly forming Christian Church to see God's remarkable intelligence running an earth-centered universe, but it also gave rise to astrology, an industry that started by using Ptolemaic mathematics to predict the positions of the planets on a person's birthday, and ultimately, to predict his or her future.

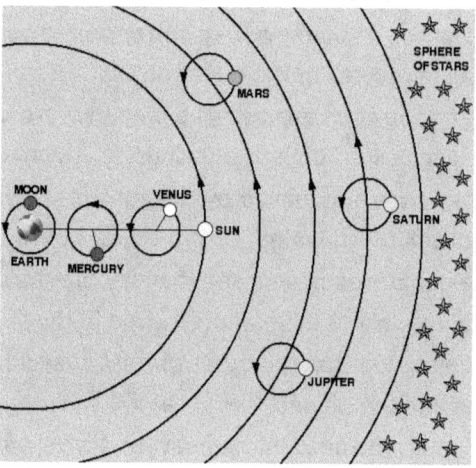

The Ptolemaic system with all of its epicycles was mathematically unwieldy and it was eventually overthrown by the much simpler heliocentric Copernican system, much to the consternation of the Roman Catholic Church. Heliocentric means a system in which the planets revolve around the sun rather than around the earth. This meant that not only the earth, but mankind itself was no longer at the center of God's universe. This displacement of the earth and its inhabitants from the center of the universe to its periphery is called the **Copernican Principle**. The church didn't like it, but it suited the mindset of the intellectual elite which was beginning to rebel against church doctrine. In addition, the removal of the superfluous epicycles meant simpler and more elegant mathematics which suited both the astronomers and the astrologers.

The expanding Universe

Skipping ahead to the early 20th century, new powerful telescopes came online, and astronomers realized that the sun was only a star, very much like other stars. Furthermore, the stars were arranged in a spiral configuration they called a galaxy. The intellectual elite were thrilled. Now man was not only removed from the center of the solar system, he was now on an insignificant planet orbiting an insignificant star.

During these early years of modern astronomy, scientists assumed that our galaxy comprised the entire universe. That changed when astronomer Edwin Hubble discovered that there were other "universes" in the form of galaxies scattered throughout outer space. Our Milky Way galaxy was only one of many. The intellectual elite were happier still! The Copernican Principle was even more firmly ensconced. Now mankind was on an insignificant planet orbiting an insignificant star in an insignificant galaxy.

Hubble also noticed that the light from these galaxies had shifted toward the red part of the spectrum and attributed this shift to the Doppler Effect. The Doppler Effect is the stretching-out of light waves due to the fact that the light source is receding from the observer. The longer you stretch light waves, the "redder" they become. This suggested that the universe is actually in a continuous state of expansion. He published a paper on this subject in 1929, and the assumption that galaxies are all receding from us became known as "Hubble's Law".

Hubble's ability to gauge distances to the galaxies was built on the 1908 discovery by Henrietta Swan Leavitt that there is a strong direct relationship between a classical Cepheid variable's luminosity and its pulsation period. (A Cepheid is a type of pulsating star.) By finding classical Cepheid stars in other galaxies and measuring their luminosities, an astronomer could then calculate the distance to that galaxy. To be fair, the proposition that the universe was expanding had been published two years earlier by the Belgian priest and astronomer Georges Lemaître, however Hubble was the astronomer who presented actual evidence that this was the case.

The Big Bang Theory

The idea that the universe began in a great explosion was originally formulated in the late 1920's but gained traction with Hubble's discovery of the red shift phenomena. If all the galaxies are receding away from each other, then by running the script backwards, one could see that there should have been a time in the beginning in which they were all squeezed together in a microscopic point called the *cosmic singularity.* This singularity was of submicroscopic size and contained the entire mass of our universe. The explosion of this singularity in the event known as "the big bang" produced our universe with all the galaxies receding from each other like we see today.

The concept of a "singularity" is quite consistent with Einstein's theories of relativity, however, in quantum theory, there is no such thing. According to quantum theory, nothing can have infinite density or infinite energy, and this fact is partly responsible for the difficulties that scientists have encountered in their attempts to unify quantum theory with relativity. Furthermore, there is no theoretical reason that a singularity, especially one that contained all the matter, space and time we see in our universe today should explode in the first place. While Einstein's theory allows for singularities with infinite mass, it does not predict that any of them should explode. Neither does quantum theory. Quantum mathematics simply stops working whenever it encounters any form of infinity, and most physicists will admit that when they encounter an infinity, nature is trying to tell them that they have come to the dead end of a blind alley!

Prior to the early 60's, the Big Bang scenario was only one among several other theories, the most prominent being *the Steady State theory* which postulates that the expansion of the universe happens because new matter is continually created. Physicist Fred Hoyle, who understood quantum physics and never believed that the universe began in a great explosion was the one who mockingly named the explosion *the Big Bang*.

In spite of the mathematical uncertainties, academic scientists decided to forge ahead. The singularity was really the only way forward, and no one gets research grants for finding new ways for covering the same ground. To this end, they created new forms of mathematics including string theory and M-theory which present their own problems and are discussed later in this chapter.

The Cosmic Microwave Background radiation

Big Bang research got a Big Boost in 1964 when a pair of radio astronomers named Arno Penzias and Robert Wilson discovered a very "cold" microwave radiation which permeates the entire universe. This radiation came to be called the *Cosmic Microwave Background* (CMB). The CMB is assumed to be the echo of the original explosion that created the universe, including matter, energy, space and time.

Using data from advanced satellites, astronomers were able to draw a map of the CMB, a sort of baby picture of the entire universe as it was when it was about 380,000 years old. Prior to that time the universe was filled with a dense, opaque "fog" of elementary particles and heat energy. At about 380,000 years after the explosion, the universe had cooled down and the opaque fog started clear and the particles that composed the fog "crystalized" into matter. Electrons began to combine with protons forming mostly hydrogen and helium, and the universe turned transparent. This clearing of the fog allowed the heat energy to radiate throughout the expanding universe unimpeded. At the time the primordial fog first cleared, the radiation was in the visible light range at about 3000 degrees Kelvin which glows a bright orange-yellow, but since that time, this background radiation has cooled to about 2.7 degrees Kelvin which is in the microwave range. It is this now-diluted heat energy that became the Cosmic Microwave Background first recorded by Penzias and Wilson.

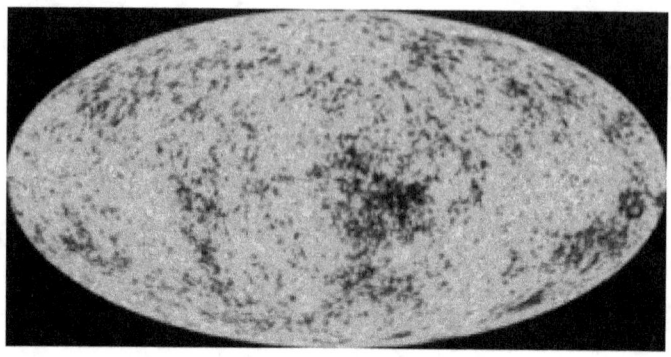

Cosmic Background map of the universe

According to the theory, the Big Bang was the event in which all of space, time and matter came into existence. Neither space, time nor matter existed prior to the Big Bang. Furthermore, the expansion of the universe is *not* caused by the galaxies rushing away from each other. It is caused by the expansion of the space between them. This is no small point. Space is no longer just an empty void. Space is an active player in the history of the cosmos and takes top billing along with matter and energy.

> *Note: Time did not exist before the Big Bang. It's difficult to imagine anything without time, but before the creation of the universe, there wasn't anything. There was nothing! NOthing! Trying to describe time prior to the Big Bang would be like trying to describe a place north of the North Pole. Space and time were created simultaneously with*

matter and energy and are bound together into
what cosmologists call "space-time".

It is important to note here that even though the
relativity and quantum theories are modeled on
mathematics, those mathematics have drawn their
reality from actual observations, and both theories have
made predictions that later proved accurate.
Mathematical models can be useful for helping to
discover real physical properties and objects which can
later be confirmed by observation, but just because a
mathematical formula predicts a physical object, it
doesn't necessarily mean that that object actually
exists. Scientific grants are awarded to researchers who
can produce progress, and if unsubstantiated
theoretical speculation is the only progress in town,
then research can easily veer away from reality and into
fantasy. In the following sections of this chapter, you
will see that Cosmology may well have found itself
sliding from a discipline based on science into
something more resembling theology.

Incidentally, modern cosmology says that all the
energy contained in the cosmic singularity was created
out of NOTHING. This surprising assumption comes
from the fact that *gravity is treated as negative energy*
which exactly balances all of the positive energy in the
universe such as heat, light, and mass. (Remember that
mass and energy are freely convertible according to
Einstein's theory of relativity -- $E=Mc^2$). Thus, if your

kindergartner asks you how much the universe weighs, you can argue that according to theoretical astrophysicists, the total mass in the universe is zero and thus would weigh nothing if measured using an earth based scale. This is exactly the sort of reasoning used in the science fiction comedy "The Hitchhiker's guide to the Galaxy", a book I highly recommend for anyone with a sense of humor!

> "It is known that there are an infinite number of worlds, simply because there is an infinite amount of space for them to be in. However, not every one of them is inhabited. Therefore, there must be a finite number of inhabited worlds. Any finite number divided by infinity is as near to nothing as makes no odds, so the average population of all the planets in the Universe can be said to be zero. From this it follows that the population of the whole Universe is also zero, and that any people you may meet from time to time are merely the products of a deranged imagination."

> The Ultimate Hitchhiker's Guide to the Galaxy, p 244 Kindle edition, copyright © 1979-1992 by Douglas Adams)

The laws of physics and the Free Parameter fundamental constants

One of the enduring mysteries about our material universe has been the discovery that virtually every aspect of it can be described by a multitude of mathematical formulas which define the laws of physics. For example, Newton's law of universal gravitation states that every point mass in the universe attracts every other point mass with a force that is directly proportional to the product of their masses, and inversely proportional to the square of the distance between them. The mathematical formula for gravitational force is $F=G(M_1M_2/R^2)$ where G is the gravitational constant and R^2 is the square of the distance between Mass 1 and Mass 2.

No scientist has ever tried to explain where this law came from. It is assumed that it has existed since the beginning of the universe. However, the question of why the universe happens to be based on coherent mathematical laws like this is an enduring question. After all, *if* the universe was based on random chance events like most scientists believe, what prevented it from evolving into a chaotic jumble of waves, particles and energy with no particular rhyme or reason?

The gravitational constant in the above formula is also something of a mystery. It has a specific value $(6.674\times10^{-11} \text{ m}^3\cdot\text{kg}^{-1}\cdot\text{s}^{-2})$, but exactly why it has this value is still a mystery. The gravitational constant is only one of 26 other fundamental physical constants whose values seem to have been arbitrarily set.

There are virtually hundreds of physical constants. You can see a list of them by clicking here. The majority of them can be derived from the mathematics of either the theory of relativity or quantum mechanics. 26 of them, however, cannot be derived from existing mathematical theory. Their values have to be determined experimentally, and they all seem to have been set purely randomly in the first few trillionths of a second after the Big Bang. These 26 fundamental constants are called "Free Parameters". When expressed in Plank units, they are dimensionless numbers. About half of them are the masses of the fifteen fundamental quantum particles (fermions), and the rest pertain to the properties associated with them.

The 26 free parameters are thought to have resulted from the breaking of unstable symmetries as the universe cooled down during the first few trillionths of a second after the Big Bang. Think of a symmetry like a pencil balanced on its point. While it is balanced, the force of gravity pulls on all sides of the pencil evenly, and for a split second, the pencil remains upright in a perfectly symmetrical state. However, almost immediately it falls over on its side. Even though the forces on a perfectly balanced pencil are symmetrical, the symmetry is unstable and the pencil falls over into an arbitrary but stable horizontal position. This is called a "broken symmetry". The pencil could have fallen over in any direction, but of course, it falls in only one. Thus the symmetry is broken and the pencil becomes "frozen" in a stable, but arbitrary state.

Physicists think that this is exactly why the free parameter constants happen to have the values that they do. The extreme heat of the early universe could maintain an exact balance of all the forces and masses, but as the universe quickly cooled off, the unstable symmetries were broken and the 26 free parameter fundamental constants were frozen into arbitrary states. Like the pencil, there were millions of physical states that any given fundamental constant could have "fallen into", but, also like the pencil, each one became frozen in a single arbitrary one. If the conditions at the beginning of our universe were even slightly different, the values of these constants would certainly have been vastly different. Physicists often "play God" by creating mathematical models of universes with bizarre physical properties based on varying the values of the free parameter constants, but none of these putative models, if they could be brought into existence, would be compatible with life.

It is a fact that if any of the free parameters were even slightly different, life in this universe would have been very different or more likely, would not even have been possible at all. Even matter might never have come into existence. At the creation of the universe, during the split second before quarks, photons and electrons came into existence (10^{-12} seconds after the Big Bang), the exact conditions had to be incredibly finely tuned in order to produce a universe in which we could live.

As a single example, if the *cosmological constant* (explained in more detail later) was different by only a few parts in 10^{120}, our universe would have either expanded too fast for matter to form into stars and galaxies, or it would have re-collapsed into a singularity within a few million years. In other words, there would be only one chance in 10^{120} that life even could have had a chance to develop in our universe. (A "google" is only 10^{100}.) As an analogy, the chances of having a cosmological constant compatible with life are less than the chances of arbitrarily picking up one specific grain of sand while rummaging through 10^{30} (an octillion) universes the size of our own observable universe.

The Anthropic Principle

The *anthropic principle* is the philosophical consideration that observations of the physical universe must be compatible with the conscious life that observes it. So far as we know, the universe we occupy, with its highly improbable combination of physical constants, is the only one that could have produced animals that had the potential to develop intelligence, and any other universe with any other combination of physical constants would be incapable of sustaining intelligent life (or more probably, any life at all!).

For example, if the *gravitational* constant were stronger than it is, the universe would have collapsed almost immediately after the Big Bang. On the other hand, if the gravitational constant were much weaker than it is, the planets and their stars might never have gravitationally coalesced from the otherwise dispersed dust and gas of the early universe.

Electromagnetism is the force that keeps the atoms in matter together. The fact that your body remains together as a coherent bundle of organs instead of dissolving into a cloud of separate gas atoms is due to the electromagnetic "glue" that holds it together. If the charge of an electron were stronger or weaker than it is, matter as we know it would have been grotesquely different, or not even possible at all.

The fine structure constant is an especially important and unusual dimensionless number associated with the development of life in the universe. It is important because if it were only 4% larger or smaller than it is, then stars would not be able to sustain the nuclear reactions that synthesize carbon and oxygen. Carbon based life in the universe would be impossible.

The fine structure constant is also unusual because it

is not even a primary constant, but a composite resulting from a "chance" resonance between four other universal constants. Nobel laureate physicist Richard Feynman called it a "magic number." In his book The Strange Theory of Light and Matter (Princeton University Press, 1985) he said:

> "It has been a mystery ever since it was discovered more than fifty years ago, and all good theoretical physicists put this number up on their wall and worry about it. Immediately you would like to know where this number for a coupling comes from: is it related to π or perhaps to the base of natural logarithms? Nobody knows. It's one of the greatest damn mysteries of physics: a magic number that comes to us with no understanding by man. You might say the 'hand of God' wrote that number, and 'we don't know how He pushed his pencil."

The *cosmological constant,* mentioned above, is another free parameter fundamental constant. It is thought to be the measure of the mysterious driving force behind dark energy, the recently discovered energy that seems to be causing the accelerating expansion of the universe. The cosmological constant equates to the energy of the empty vacuum and it is a mystery why its value is so close to zero (10^{-122} plank units = approx. 10^{-178} Newton-meters. Earth's gravitational pull on a kilogram of mass is 9.8 Newtons.) If the cosmological constant were even the tiniest bit larger, the expansion of the universe would have been so fast that stars, planets and galaxies would never have formed, and if it were any smaller, the universe might have collapsed within a few centuries of the Big Bang. Our existence on earth is dependent on the exact value of this tiny fundamental constant. Dark energy is discussed in more depth later in this chapter.

The question of how an anthropic universe ever came into existence against such low odds (much, much less than $1/10^{120}$) has vexed some very clever scientists. The astrophysicist Fred Hoyle said that he thought that the universe was a "put up job" and likened the probability that random chance could account for our anthropic universe was about the same as the chance that a tornado could whip through a junkyard and assemble a 747 passenger jet

Inflation

Except for the singularity itself, the Big Bang theory was all very mathematically tight, and everything worked perfectly...until the astronomers began to make more careful observations of the actual universe. Armed with Einstein's theory of relativity, more powerful telescopes, computers and the newly discovered CMB, a number of phenomena were discovered which did not fit into the original theory, and it became increasingly necessary to add various postulates to make the theory align with the observations.

Explosions are generally very messy affairs as anyone who has seen one can attest. They are thoroughly chaotic and hurl gas, dust and clumps of debris unevenly in all directions. The gigantic explosion that began the universe *should* have been no exception. It should have produced a universe that was uneven, clumpy and as messy as a pile of garbage...But observation reveals a very smooth universe with galaxies nearly evenly distributed throughout space. Telescopes reveal a universe in which every sector looks pretty much like every other sector (heterogeneous), and one that behaves the same in every direction (isotropic). The CMB map was at first thought to be entirely smooth but it revealed extremely tiny density variations only after turning up the contrast.

In the early 1980's an American theoretical physicist named Alan Guth came up with the ultimate postulate to explain away many of the faults in the original Big Bang theory. **Inflation** explains why the universe appears to be the same in all directions, why the cosmic microwave background radiation is nearly uniform in every direction, why the universe exhibits flat geometry, and why no magnetic monopoles have been observed.

Inflation is a phenomenon believed to have

happened at the birth of the universe in the initial stages of the Big Bang. It happened between 10^{-36} and 10^{-32} seconds after the initial "explosion". During this incredibly tiny period of time (a few billionths of a trillionth of a trillionth of a second), space expanded from the size of a proton to about the size of a grapefruit at a speed much faster than the speed of light. Theoretically, this *is* possible, even though Einstein stated that nothing can travel faster than the speed of light. There really was no matter at this early epoch (just enormous heat energy), and Einstein's speed limit only applies to physical matter moving through space...but not to space itself.

This rapid expansion of the universe smoothed out the irregularities inherent in the original explosion just like inflating a wrinkled-up balloon smooths out its wrinkles. To extend the analogy, the more you inflate the balloon, the flatter its surface becomes. In the same way, the early hyper-spherical universe began to assume a flatter geometry. (The term "flat" refers to a hyperspacial fourth dimension in which the universe could be observed from a God's eye perspective outside of time and space. It looks like hyper plane, but it *could* actually be a hyper sphere or a saddle shaped open hyperbola.) Although there is some disagreement about the degree of flatness of the universe, most astrophysicists agree that it is at least a "nearly" flat hyper plane (to over a 99% confidence level).

Unfortunately, the phenomenon of inflation is a mathematical invention with no observable characteristics or measureable fundamental constants. Not only has it never been observed, but it is unlikely ever to be observed. Cosmic inflation has become not only the savior of the Big Bang theory of the universe, but also the corner stone upon which the entire Church of Cosmology has been founded. No mechanism has ever been found (or even theorized) that could account for cosmic inflation.

Theoretical physicists assume that a currently unknown *scalar field* that was dominant in the very early phase of the Big Bang might account for cosmic inflation.

(*Note: The term scalar is a property defined by a single number like speed, volume, mass, temperature, power, energy, and time. Other fields are defined by vectors. A vector is defined by two numbers. For example in the case of a magnetic field, the magnetic lines of force are defined by direction and force. Velocity, another vector is defined by speed and direction.*)

It was hoped that the recent discovery of the Higgs Boson and its associated field could solve the mystery of inflation since it is the first scalar field ever discovered in quantum mechanics. So far, however, it does not appear to have the anti-gravity characteristics that can account for cosmic inflation.

After the invention of cosmic inflation, this hypothesized phenomenon has now been expanded (no pun intended) to explain nearly everything else that the original Big Bang theory could not explain, including the incredible fine tuning of the fundamental physical constants discussed earlier in this chapter. As you are about to discover, modern cosmology now assumes that an eternal inflation field is the god that created not just our universe, but an infinite number of other universes as well.

Dark Matter and Dark Energy

In spite of the panacea of inflation, a number of new observations that no one expected have cropped up. These have necessitated the invention of new tweaks to the Big Bang theory. They include **dark matter**, which is believed to constitute 27% of the mass of the universe, and **dark energy**, which constitutes 68% of the mass of the universe. This leaves ordinary baryonic matter (the stars, planets, dust, gas, life…etc.) to comprise only 5% of the mass of the universe. Apparently mankind has spent thousands of years studying only 5% of our universe. We still have no idea about the nature of the other 95%.

Dark Matter

Dark matter was first postulated when it was discovered that observable galaxies were lacking enough mass to retain their shapes. Later, it was also discovered that there was not enough observable matter in the universe to make it as "flat" as it appears to be. In order to have a flat universe, about 27% more matter than can be seen using all available resources would be necessary. Hence, another tweak was added to the theory in the form of a mysterious and possibly exotic type of matter who's only known property is the gravity it exerts. The theories that account for the nature of dark matter fall into two categories.

First there are the MACHO's. This is an acronym for Massive Astrophysical Compact Halo Objects. These include all the baryonic (normal matter) candidates for the less visible astronomic bodies such as black holes, brown dwarf stars, neutron stars and small white dwarf stars, not to mention asteroids, dust, interstellar gas and planets. Unfortunately, MACHOs are not likely to account for the large amounts of dark matter currently thought to be present in the universe because calculations of the amount of baryonic matter produced during the big bang (and other considerations) have ruled them out.

The second candidate are the WIMP's. WIMP is an acronym that stands for Weakly Interacting Massive

Particles. WIMP's are theoretical elementary particles that interact only weakly or not at all with ordinary baryonic matter. All efforts to detect any type of WIMP have been failures. The theory predicting these particles springs from the theory of *Supersymmetry* which was formulated to solve a slew of existing puzzles inherent in the standard model, especially the contradictions involved in unifying the strong force with the electroweak force.

Supersymmetry hypothesizes that every subatomic particle has a supersymmetrical partner. These particles bear the same names as the original quantum particles, but are each preceded by the letter "s". Thus there is the *squark*, the *selectron*, the *sneutrino*...etc. There are also such supersymmetric zingers as the *photino*, the *wino*, and the *gluino*. All calculations suggested that these ghostly particles should materialize at energies compatible with the new LHC (Large Hadron Collider). Unfortunately, not one of these super symmetric partners has ever been found, largely disproving the entire theory of Supersymmetry.

Dark Energy

Dark energy became an issue because in 1998 the Hubble Space Telescope discovered that the universe today is actually expanding faster than it was just a few billion years after the Big Bang. Prior to the 1998 discovery, it was assumed that the expansion of the universe was due to the inertia imparted to matter during the big bang explosion, and that the cumulative gravity of the matter created in the original Big Bang would gradually slow down the rate of expansion of the universe. Instead, the galaxies were observed to be receding from each other at an *accelerating* rate. No one really knows how to address this problem definitively, and it is no small problem. Cosmologists can only assume that some form of dark energy must be responsible, and in order to produce the acceleration observed through modern telescopes, this mysterious form of energy must constitute about 68% of the mass of the universe. The astrophysicists cannot settle on what it is or its mechanism of action. At the moment, the only thing that they can put a finger on is the energy that defines the *cosmological constant*.

To clarify this mysterious expansion of the universe, it is important to point out that:

- The expansion applies to all the matter in the universe that is too far apart to be gravitationally bound.

- The matter itself is not "exploding apart"as a result of inertia like the debris in a bomb explosion. Rather, the expansion is due to the mysterious expansion of the SPACE between material objects. Space itself is expanding, and the objects within it moving apart because of it. Inertia has nothing to do with it. This is analogous to dots on the surface of a deflated balloon expanding apart when the balloon is inflated.

- The further apart two objects are, the faster they will move away from each other.

- The universe is self contained and there is no such thing as space or time outside of its boundaries. Therefore there is no "space" for the space of the universe to expand into, in the same way that there is nothing north of the north pole.

- The expansion began to accelerate about four billion years ago when the matter in the universe had spread so far apart that a majority of it had become gravitationally unbound.

The cosmological constant

The cosmological constant was originally a simple variable in the theory of general relativity which Einstein used as a "fudge factor" and which he later disavowed. He had no idea about how the universe came into existence. He assumed, just like everyone else at the time that the stars and planets had existed in a stable state since the beginning of time. Unfortunately, his original mathematics predicted that the universe was unstable and that the cumulative gravity of all the matter in the universe should eventually cause it to begin a collapse.

Einstein believed that the universe was in a never changing steady state, so he needed a factor to remove this unwanted instability in his calculations. He therefore added a cosmological constant lambda (λ) as a theoretical anti-gravity force to counteract gravity and allow the universe to remain in a steady state. The cosmological constant was the measure of an unknown and unwanted anti-gravity force that could counteract the natural tendency of gravity to collapse the universe.

This was a rational assumption in 1917 when he formulated his theory. Everyone at that time believed that the universe was eternally static. Much to Einstein's chagrin, however, when astronomer Edwin Hubble discovered that the galaxies were all receding from each other as a result of inertia from what eventually became known as the Big Bang, he realized that he had made a mistake, and in 1931 he removed the term from his equations saying that it was his "greatest blunder".

The assumption after that point was that the universe would either expand forever as a result of the inertia of the Big Bang, eventually reverse its expansion and begin to contract as a result of the constant pull of gravity, or reach a steady state in which the gravitational contraction exactly matched the inertial expansion. The outcome of that question would depend on future measurements of the rate of expansion. Note that in 1931, the expansion was assumed to be due to the inertia imparted to matter during the initial big bang. The realization that the expansion of the universe is really due to the expansion of the SPACE *between* the galaxies came much later.

Scientists didn't give it much thought after that. They assumed that the value of the cosmological constant was equal to zero until 1998 with the discovery that the universe was not just expanding, but expanding at an accelerating rate. After this discovery, astronomers began the process of objectively measuring its value using their telescopes. At the same time theorists began trying to derive its value using mathematics and the recent discoveries of quantum physics.

The *observed* measure of the cosmological constant's value is approximately 2.888×10^{-122} in Plank units. Notice the negative exponent. You can see that the cosmological constant is so close to zero that it almost doesn't exist at all. It was a surprise because suggesting to astrophysicists that such a tiny force acting over the full extent of the universe could have such a large effect would be like suggesting that if I gained five pounds, it would destabilize the orbit of the earth. (So far, so good!)

Unfortunately, the *theoretical* value of the cosmological constant turns out to be some 120 orders of magnitude greater than the observed value. OOPS! The discrepancy between the theoretical and the observed values for the cosmological constant is a problem for theoretical physics. This disagreement is called the *vacuum catastrophe.* It is an example of where the gap between theory and observation was so large that mathematical theory had to be abandoned entirely. Since its true value cannot be derived by theoretical means, the cosmological constant is therefore considered to be a fundamental physical constant (a free parameter baked into the universe at its birth) and no one has any idea about why it exists or the true nature of the energy that it describes!

The cosmological constant does not "explain" the accelerating rate of the expansion of the universe because it is only a measured value. It is not itself a form of energy! It doesn't really explain *why* the universe expands at an accelerating rate. It simply tells you *by* how much. The "why" now stands as a tentative hypothesis grounded in the energy of the vacuum.

Quantum theory postulates that what we call empty space is not really empty at all. Subatomic particles, called "virtual particles", are always forming from nothing and annihilating nearly instantaneously. While the total mass and energy of all this new "matter" adds up to zero (because the sum of creation and immediate annihilation adds up to zero), the virtual pressure that it exerts within empty space, while tiny, adds up over the immense volume of space to create enough "virtual energy" to total 68% of the entire mass of the universe.

No one knows if this is the correct explanation for dark energy, but it is the best explanation we have so far.

The cosmological constant has become both a mystery and one of the most important and fine-tuned fundamental constants. If it were the tiniest bit larger (on the order of 10^{-122}), the expansion of the universe would have been so fast that stars, planets and galaxies would never have formed, and if it were any smaller, the universe might have collapsed within a few centuries of the Big Bang.

4
The Many Worlds interpretations

Hugh Everett was a student at Princeton University in the mid-1950s. One night in 1954, he and a couple of his classmates were sitting around drinking and, in his own words, thinking up "ridiculous things about the implications of quantum mechanics." In the quantum world, a quantum object like an electron, for example, can exist in a superposition of locations, velocities and spins. As we have seen in chapter 2, we never actually see the electron in a superposition state. Nor do we see elementary particles or macroscopic objects in superposition states.

Everett's idea was to ask "What if the Schrödinger equation always applies to everything—objects and observers alike? What if the superpositions never collapse, and even *people* remain in superposition states?" He may have been drunk when he formulated the theory, but his genius was that he could model this possibility using mathematics (presumably after the hangover), and more importantly, his model was the first, and ultimately the only one that removed the Copenhagen Interpretation's otherwise messy requirement for conscious observers.

Under this assumption, the wave function of an observer would bifurcate at each interaction of the observer with a superposed object. Each branch has its own copy of the observer as well as the observed object. Each copy of the observer would have seen a different outcome to his experiment. *Each copy occupies its own universe!*

In this interpretation, the observer can be eliminated as causing a collapse of the wave function because every outcome comes to fruition with or without a conscious observer. It is a sort of ultimate "Murphy's law". Everything that can happen does happen within an infinite tree of branching universes. In one multitude of universes, you are reading this book, but in another multitude of universes, you died last year in a plane crash. In some universes, you might have a pet dinosaur. You, and everyone else in the world exists simultaneously in an infinite variety of circumstances, and in some universes, life never even developed on earth! Everett wrote: "From the viewpoint of the theory, all elements of a superposition (all 'branches') are 'actual,' none any more 'real' than the rest."

Bear in mind throughout the following discussion of multiverses that *Hugh Everett's concept of infinitely multiplying universes is THE ONLY ONE that offers an alternative interpretation of quantum physics that entirely bypasses the necessity for a conscious observer to collapse Schrodinger waveforms thus creating real particles.* It's the only quantum theory that allows for a multiverse without consciousness.

As wacky as this theory may sound, it was the world's first foray into what would eventually evolve into the various "multiverse" theories, and a coterie of physicists embrace Everett's Interpretation of quantum mechanics as an alternative to the Copenhagen interpretation.

David Deutsch, a founder of the field of quantum computation which was inspired by Everett's ideas had this to say about it.

> "He (Everett) represents the refusal to relinquish objective explanation. A great deal of harm was done to progress in both physics and philosophy by the abdication of the original purpose of those fields: (By this, he means a hard scientific philosophy that does not include consciousness as a necessary component of the universe.) We got irretrievably bogged down in formalisms, and things were regarded as progress which are not explanatory, and the vacuum was filled by mysticism and religion and every kind of rubbish. Everett is important because he stood out against it."

"Mysticism, religion and every kind of rubbish!" Where to begin? Physicists fall in love with any mathematical formalism that can accurately paint a detailed picture of a reality that can be used to advantage in our own. But a picture, no matter how detailed and beautiful is not reality. The mathematical picture of Everett's many worlds vision might be a useful tool in the calculation of other theoretical mathematical models, but if you try to apply it to our conscious understanding of reality, it boggles the mind and makes absolutely no sense.

According to Everett, every time a random event of any kind happens, even the random decay of a radioactive atom on a planet on the other side of the universe, it instantaneously creates another slew of universes in which you and everyone else in the world is duplicated and careens off into an entirely different future. Furthermore, if Everett's parallel universes actually exist, we have no way of ever detecting them or proving their existence. Moreover, not a single one of those parallel universes can ever have an effect on our own (except as a plot in The Twilight Zone).

Everett's Many Worlds Interpretation was the very first step towards the Multiverse theories espoused by many scientists today. The multiverse is a concept tenaciously held by modern scientists and secular humanists (atheists) alike. It presents a scenario for the creation of our universe, as well as countless other universes without assuming that there was a consciousness that most of us call God at the root of it all.

Multiverses

The theoretical phenomenon of inflation opened up a virtual Pandora's Box of mathematical possibilities, the most significant of which is the concept of an unlimited, number of other "parallel" universes that exist alongside our own. An important aspect of most of these parallel universe theories is that our universe, as well as all the parallel universes exist in a hypothetical **hyperspace**.

According to the Eternal Inflation theory, the *larger* "multiverse" can be thought of as a wheal of Swiss cheese with an eternally existing inflationary field being the cheese and multiple universes being the holes in the cheese. This Swiss cheese expands forever at a rate much faster than the speed of light creating an infinite number of "holes" (meaning new universes) with no known or even hypothesized beginning and no known end. Nor does this cheese (hyperspace) have any definable properties other than being an infinitely energetic and ever expanding field. The inflationary field is entirely hypothetical, and is likely to remain that way since it lies outside of our universe and can never be observed directly.

A considerable number of very prominent astronomers and cosmologists dispute the cosmic inflation theory and lament the fact that the theory is intrinsically untestable and therefore "unscientific". But undaunted, the cosmologists keep trucking on and each time a new cosmic observation comes to the fore, various scientists find a way to fit it into the theory using newly added mathematical tweaks.

SuperString Theory

Neither quantum theory nor the theories of relativity have been able to account for all the properties of the universe. What became known as the *standard model* beginning in the early 1960's was an attempt to mathematically describe the entire universe in a single theory using standard forms of mathematics. The standard model incorporates both the theory of relativity and quantum theory to integrate matter and energy in the universe. This includes a mathematical description of the hadrons (composed of quarks), the leptons (which include electrons and neutrinos) and the bosons (which include the photon and carry the four known atomic forces).

The Standard model predicted the existence of twelve theoretical force-carrying particles called bosons (photons are the most familiar example) and the Higgs particle prior to their actual discovery using particle accelerators. The standard model also predicted the anti-matter counterparts to the hadrons and leptons of normal matter as well as unusual short-lived combinations of quarks such as mesons and pions. The standard model was also reasonably successful at *unifying* three of the four fundamental forces.

Unfortunately, there are several major deficiencies in the standard model. The unification of the strong nuclear force with the electro-weak force is haphazard, gravity has never been integrated, and the standard model does not account for the arbitrary values of the 26 free parameter universal constants. In addition, the standard model has not been able to explain the dark matter or dark energy which have recently been discovered and found to be extremely important components of the universe.

Thus the standard model gives the impression of not being entirely finished. This prompted theoretical physicists to search for a more comprehensive mathematical treatment that could remedy these shortcomings. Such a theory would be called a "No-Free-Parameter Grand Unified Theory" (GUT), or a Theory of Everything (TOE). This goal has become the holy grail of modern physics.

Superstring theory is one such attempt and is an alternative to quantum theory. As you recall from our discussion of Einstein's theory of relativity in chapter 1, Einstein (and the authors of quantum theory) used standard mathematics to formulate their theories, and their theories have proven over time to predict real world phenomena. Superstring theory, however is an entirely new form of mathematics. Although it CAN be made through various tweaks to predict real world phenomena, it also predicts phenomena that lie entirely outside of scientific comprehension. The tweaks, of course, derive from attempts to make the theory fit real-world observations. Keep this fact in mind as you wade through the following discussion of superstring theory and the cosmology based on it.

Superstring theory visualizes sub-atomic particles as tiny one-dimensional strings which vibrate in *ten spatial dimensions*. We, of course, live in a three dimensional world (length, width and height) plus a fourth time dimension. Scientists call our four dimensional reality *"space-time"*. Virtually all of the mathematical evidence for a ten dimensional space-time with cosmic inflation and multiverses is entirely hypothetical and contained in *superstring theory*.

"Strings" are the hypothesized higher dimensional counterparts to the point particles (hadrons and leptons) in the standard quantum model. The strings are incredibly tiny (on the order of a Plank length, 1.6×10^{-35} meters), and they vibrate in different harmonic resonances, like the strings on a guitar. The different harmonic resonances define the type of particle any given string appears to be (electron, positron, quark, neutrino, photon...etc.).

Unfortunately, Superstring theory can only be made to "predict" some of the known quantum entities by force feeding it specific values that were originally derived from the standard quantum mechanical theory.

Furthermore, there are virtually an infinite number of "tweaks" that can be applied to the theory, and there are no physical laws or conditions that can definitively determine which tweak(s) to prefer over the others. Over time, the physics community settled on five different versions of superstring theory, all of them different and none favored over the others.

The famous theoretical physicist Ed Witten, after much labor, discovered that by adding one additional dimension he could (mathematically) unify the five different versions of superstring theory. What was once known as superstring theory then morphed into a much more complex mathematical structure called **M-theory** which operates in 11 dimensions rather than the original 10 of superstring theory. In order to make this successor to superstring theory work properly, it became necessary to postulate that instead of only one dimensional strings, you needed, at minimum, two dimensional membranes. Think of very thin rubber sheets. If you roll up a sheet into a tight straw, it looks just like a string, so strings and membranes can exist side by side. In addition, these membranes need to bend and vibrate in no less than 11 spatial dimensions; seven more than our familiar four dimensional world.

Where are these seven extra dimensions? Theoretical physicists tell us that they are directly in front of us. They say the extra dimensions are very tiny and rolled up, so they are out of sight. This is referred to as "compactification". According to the theory, the extra dimensions are compacted in extremely tiny 6 dimensional shapes called Calabi-Yau manifolds, each one related to a different "vacuum state". Unfortunately, there are an enormous number of these shapes (10^{500} is the assumed number because of calculations involving "brane fields"—explained below). Each Calabi-Yau manifold predicts a different possible universe with different characteristics, and there is no way to discern which of these shapes may be relevant to *our* universe. There is no way to verify that these compacted dimensions actually exist other than the enormously complex mathematics that predicts them. So far, there are no experimental results that have indicated that the extra seven dimensions are anything other than a mathematical fiction.

M-theory also infers that there can be other higher dimensional *"branes"* in the hyperspacial vacuum. These Branes are referred to using numbers such as 0-branes (points in space), 1-branes (strings) 2-branes (two dimensional rubber sheets), 3-branes (rubber spheres, cubes and other 3 dimensional shapes) 4-branes (unimaginable 4 dimensional shapes including hyper-spheres and hyper-cubes sometimes called tesseracts) all the way up to 10-branes, all of them vibrating in any number of the 11 postulated dimensions. Each "brane" creates a "flux" around itself analogous to the electric field surrounding an electron. M-theory calculations suggest that there are 10^{500}

possible flux fields, each associated with a differently shaped Calabi-Yau manifold, and thus, 10^{500} entirely different *types* of universes.

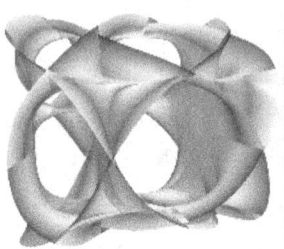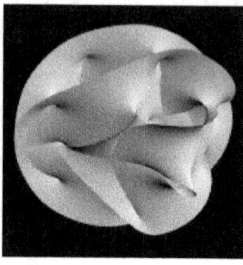

Two examples of 2D slices of 6D Calabi-Yau quintic manifolds.

This has opened up a cornucopia of fanciful speculation for cosmologists. Some have postulated that our universe exists on one membrane surrounded by other universes on the same brane, which is itself immersed in an unimaginable (possibly infinite) number of other parallel branes each home to other parallel universes. This entire multiplicity of branes with their parallel universes exists in a "bulk hyperspace" that continues to expand forever, probably producing more and more branes as it expands. Exactly where this monstrous hyperspace exists is not divulged by the mathematics. Nor are any of its properties. Presumably it is of infinite extent and has existed forever, outside of time itself.

Science, Math and God

In his excellent book <u>Not Even Wrong</u>, mathematician Peter Woit says:

> The possible existence of, say, 10^{500} consistent different vacuum states for superstring theory probably destroys the hope of using the theory to predict anything. If one picks among this large set just those states whose properties agree with present experimental observations, it is likely there still will be such a large number of these that one can get just about whatever value one wants for the results of any new observation.

On the other hand, without string theory there would be no rational for an infinite number of universes, and *without parallel universes, our universe would be* ***"special"***. It would be a tremendous violation of the Copernican Principle, and that's something that is unthinkable in our politically correct university science departments.

But **Quantum** theory (as well as relativity theory) is based entirely on tenured classical mathematics. It also springs forth from real-world phenomena that have actually been observed. Once formulated, the math that describes quantum theory has been used to predict other mystifying properties such as non-locality and myriad other aspects of the sub atomic world. Experimental evidence has proven that all of its predictions are valid. Without quantum theory, we would never have developed advanced electronics, lasers, computers or that apex of human civilization, the IPhone. Quantum theory has stood the test of time. The same cannot be said of string theory or M-theory.

No one has ever found any physical or even theoretical evidence of either the bulk hyperspace or its multitude of parallel universes outside of M-theory. So far, the multiverse and its associated superstring theories are really just the result of elegant mathematics.

There is no question that two unicorns plus two unicorns equals four unicorns. This does not, however prove that unicorns exist. **The multiverse is really a mathematical theology.**

Steven Hawking, was, until his death in 2018, the most famous wheelchair bound theoretical physicist in the world. He was an enthusiastic fan of M- theory and believed that it would ultimately prove to be the theory that finally unified Einstein's theories of relativity with Bohr's quantum theory. Like many other academic physicists, he believed in a multiverse consisting of 10^{500} parallel universes based on the number of Calabi-Yau manifolds hypothesized in M-theory. But Hawking was a brittle secular Humanist and one suspects that buried deeply in his subconscious at least part of his reasoning involves an attempt to avoid a metaphysical explanation for the vanishingly small probability that our universe came into existence in the first place.

10^{500} universes explains nicely the enormous improbability that our universe just happens to be the one in a "gazillion" in which all of the universal fundamental constants and laws of nature have exactly the right values and exactly the correct coherence to be compatible with a universe that can support stars, planets and life. If the chances against the random creation of our universe were the same as the random selection of an anthropic cosmological constant $(1/10^{120})$, then out of 10^{500} universes to pick from, there should be a near infinity with the same values as the cosmological constants in our own.

But 10^{500} is such an enormous number that it is impossible to contemplate. If you calculate the number of grains of sand in all the deserts and all the beaches on earth (using an average particle size), you get an estimate of **7.5×10^{18}**, or seven quintillion, five hundred quadrillion. That's a tiny, tiny number compared to 10^{500}. The best modern estimates place the number of *atoms* in the observable universe at "only" about 10^{80}. If our universe contains 10^{80} atoms, then 10^{81} atoms would add up to a total of 10 universes, 10^{82} would add up to 100 universes and it would take a trillion trillion trillion trillion... (Out to 34 trillions) of *universes* the size of the one we occupy just to contain 10^{500} *atoms.* And Hawking was talking about 10^{500} *universes*! That's a number that even dwarfs America's national debt!

How many types of multiverses are there?

Keeping in mind that there is no way that we can ever see, sense or otherwise verify the presence of universes outside of our own, scientists have theorized all together nine different scenarios for how parallel universes might be structured. All of these scenarios can be mathematically modeled, and there is no real evidence favoring the existence of any one over the others although number 1 below seems to be the most popular.

Number 1 is *the inflationary universe*. It is the "Swiss cheese" eternal inflationary model (mentioned above) which postulates a timeless, constantly expanding inflationary field in which "tiny" holes of calm form just like the holes in a Swiss cheese. Each hole equates to the cosmic singularity which initiated our "Big Bang" and each one goes on expanding into a separate universe. Each universe would have its own version of space, its own unique set of fundamental constants and possibly its own version of matter. Within these holes of relative calm, matter may or may not crystallize. The crystallization of matter would depend upon the fundamental physical constants and physical laws with which that particular universe happens to be endowed.

If the fundamental physical constants happen to align properly, elementary particles might begin to precipitate like raindrops. These might go on to form protons, neutrons electrons, stars, galaxies, planets and all the various wonders of our own universe. It stands to reason that in an infinite inflationary field, infinite numbers of universes would form making our universe only one of another infinite number of similar alien universes. And, of course, in an infinite number of similar universes, some would be exact copies of our own universe, some even containing exact copies of you reading exact copies of this book.

Number 2 would be a *"quilted multiverse"* in which there is only one volume of (ordinary) space in which other universes spontaneously form, a bit like soap bubbles in a bubble bath. Space, in this case, would be infinitely large. These universes may occupy the same space as ours, but they all exist at such vast distances and are expanding so quickly from us that they are unobservable. They lie forever outside the horizon of our visible universe.

Number 3 is the *"brane" multiverse* which is the one described above under the heading of M- theory. Each brane might contain one or an infinite number of universes and there would be an infinite number of these branes floating around in a theoretical hyperspace that looks suspiciously like the "Swiss cheese" hyperspace mentioned in number 1.

Number 4 is the *cyclic multiverse*. This is an old one that originally assumed that our universe would continue to expand, but eventually the expansion would slow down and begin contracting until it ended in another cosmic singularity. This would result in another explosion meaning another Big Bang, sometimes called the "Big Bounce". This expansion and contraction and re-expansion would happen over and over again creating a new universe to replace the older one ad-infinitum.

Number 5 is the Landscape Universe. This scenario relies upon M-theory and is very arcane. It involves a landscape dotted with multiple universes that can suddenly move (in their entirety) into different locations on the landscape through the quantum phenomenon of tunneling. Mostly, however, it is an attempt to reconcile the fundamental physical constants, especially the *cosmological constant*, with the anthropic principle. As you recall, the anthropic principle states that the universe HAD to be structured the way it is. Otherwise life would never have been able to evolve. The theory does this by manipulation of the hidden compactified dimensions (see string theory above).

Number 6 is the *quantum multiverse*. This goes back to the one Hugh Everett dreamed up while drinking with his buddies. Every time any quantum event (actually, ANY decision) occurs, instead of breaking the superposition, both states solidify and go on to create a new universe. These universes exist right next to us in our own space, but are invisible because they exist in different dimensions. This is the only multiverse which offers a scenario that avoids the need for conscious observation in order to bring quantum objects into reality.

Number 7 is the *holographic multiverse*. This one could apply to several of the above universes created in a hyperspacial field. It assumes a multiverse with a gigantic mass of "soap bubble" universes. The bubbles are defined by their *surface areas* only. They contain nothing real in their volumes. Everything contained in each of the soap bubble universes is simply a hologram projected into the interior from the image imprinted on the bubbles' surfaces.

Number 8 is the *simulated multiverse* which sees everything as an illusion created by a gigantic computer program like the one seen in the film "The Matrix" starring Keanu Reeves. Red pill anyone? Since the universe is just a computer program, any number could be running in the universal computer at the same time, hence the simulated multiverse.

The simulated universe has become (believe it or not) part of the mainstream multiverse theory, and it really *is* espoused by numerous theoretical physicists. One of the major difficulties with this theory is that if our universe is simply a computer simulation, then there is no reason to believe that our laws of physics (and everything else we have learned about our world) is real, or that the world wasn't created five minutes ago along with our memories and a history complete with dinosaurs and politicians!

Number 9 is the *ultimate multiverse*. This one assumes that all mathematically possible multiverses can and do exist. (All of the above.)

Multiverse Conclusion

Today there are three mathematical models that accurately describe the real world we live in. They are Newtonian Physics, Quantum Mechanics and Einstein's Theories of Relativity. Astrophysical theory (cosmology), on the other hand appears to have raced into a quagmire of theoretical wishful thinking based upon fanciful mathematics and contrived phenomena that have never been observed and are unlikely ever to be observed. Multiverse theory, and, even the Big Bang theory itself have become indecipherable labyrinths of mathematical mythology preached by a coterie of scientific high priests who can speak their esoteric language. They have evolved into modern creation myths, and they hold a sacred place in the doctrines of the church of secular humanism.

The famous Oxford professor Richard Dawkins (The God Delusion, The Selfish Gene, The Blind Watchmaker) regards religion to be "blind faith in the absence of evidence." He argues that "truth means hard evidence". He considers traditional creation stories to be bronze-age myths because there is no scientific evidence for them. When confronted with the argument that God created the universe, he counters "Yes, but who made God?"

On the other hand, Dawkins, like so many other scientific atheists is very willing to believe in the Big Bang and theoretical multiverses as a "rational" form of creation story. For him, it contains nothing but hard evidence! But you might want to ask him who made the branes, or the strings or the Calabi-Yau manifolds or the 11 dimensions of space? Where did the Inflationary field (Hallowed be thy name!) come from? What are its properties? Why does the universe look like "a put-up job" (in the words of astronomer Fred Hoyle?) More importantly, who created the *laws* that govern the *creation* of infinite numbers of universes out of nothing?

In fact, no one knows how the laws of nature came into being. Neither can physicists say why the laws can be described via mathematics. Dawkins lives in a theological fiction. His beliefs are even more bizarre and unbelievable than western religious belief in Moses and the burning bush.

Finally, the Big Bang Theory has become a sort of "ten commandments" upon which the entire church of cosmology rests. Unfortunately, the big bang theory has become so hopelessly frayed, not only around the edges, but throughout its entire structure that even its most ardent proponents must at least occasionally wonder if the universe actually did begin in a gigantic explosion. Cosmic inflation, a concept that is thought to have occurred between 10^{-36} and 10^{-32} seconds after the initial "explosion", and lasted for only a few billionths of a trillionth of a trillionth of a second was hypothesized to iron out several glaring inconsistencies in the original theory. While it may have served its original mathematical purpose, there is no direct evidence that it actually happened. Since its invention, other observational inconsistencies have cropped up, and have required new inventions such as dark matter and dark energy. The number of "patches" needed to mend the big bang theory has multiplied to the point of absurdity. It's starting to look as if the wider universe may not have begun when and how we have been lead to believe. Perhaps the universe really IS infinite in extent and in time...and given infinity, ANYTHING is possible. (See Boltzman brains above.)

Throughout thousands of years, literally billions of people have sensed, without scientific instruments, that an unseen spiritual world exists in a separate dimension intimately intermingled with our own. These beliefs have sprung up spontaneously throughout history in every society since people first walked on the earth. Even if the ancient mythologies seem fanciful, they are frankly no more outrageous than 10^{500} mathematical universes. Not one of them has ever needed 7 compacted dimensions to justify their gods! And the people who believed in them never needed complex machines or higher math to conclude that human consciousness is the key to understanding the universe!

These naturally occurring beliefs have another more subtle advantage denied to the atheists who rely on cosmological fantasies to justify their secular humanist beliefs. They allow us to perceive a PURPOSE to the apparently chaotic world in which we live. Who among us will be remembered by anyone alive in 200 years? Everything in our material world eventually withers and dies. Who has ever, in a time of unspeakable tragedy taken solace in the thought of an eternal inflationary field? How many people over the millennia have seen, felt or sensed any evidence of a cold multiverse in which each person is of infinitely less significance than a grain of earthly sand?

But anyone who believes in a spiritual world can take comfort in an eternally existing soul; a soul that can continue learning, loving and understanding forever; one that never needs to mourn the loss of a loved one; one that can live a life without the fear of death. Even Anthony Flew, one of the 20th century's most famous atheists converted to a belief in God before his death. As the old World War II saying goes, "There are no atheists in foxholes". The ability to acknowledge our instinctual belief that we are here for a much greater purpose can spur us to achieve greater things in our lives.

This book is part 1 of a 2 part series. The book you are reading is really a refutation of the enormous hubris displayed by the secular humanists that rule Western (and Communist civilizations). It isn't meant to be a proof that a spiritual world exists, but rather exposes modern secular scientific atheism for what it is and shows that God really DOES have a place to live. The second book in the series is entitled The Structure of Heaven. It is a good deal longer than this one and discusses the evidence for the world of Spirit. I wrote it to assure those who are ready that no one lives forever, but no one ever really dies!

5

What is consciousness and how does it work?

We know from the discussion above, that consciousness is a fundamental property of the real universe. But the question of how consciousness is created, or what it is, has never been answered in spite the enormous advances in the technology scientists have invented to study the problem. It is a rather important question for quantum physics in light of its findings about the role consciousness plays in transforming quantum reality into the reality we experience every day. From the EPR paradox, through Schrodinger's cat and Bell's theorem, *consciousness* is the key to a universe that exists in a definite state. Unfortunately (for materialists, in any case) no theorist has ever found anything within material science which defines, or even begins to explain consciousness.

There are essentially two theories that attempt to explain how consciousness relates to the brain.

The first is really a group of many **production theories.**

The production theories all assume that consciousness is **produced** by the brain itself, either as a primary product of brain wiring, or as an *epiphenomenon,* which means that consciousness is simply an artifact of brain wiring, something that began as a sort of useless side effect produced by the brain while in the process of producing the body's various reflexes and its instinctual behaviors.

The second theory is the **transmission theory**.

> The transmission theory assumes that consciousness resides outside of the body and the brain acts as a two way radio that **transmits** consciousness between itself and wherever consciousness lives. Needless to say, this theory is not accepted by skeptics, but it has some surprising support from some very influential researchers. We will be dealing with the transmission theory in the next chapter.

Production theories and their problems

The production theories assume that consciousness is "simply" a matter of neurological connections in the brain. These theories maintain the materialist view that what we perceive as consciousness is just an artifact of the brain, an "epiphenomenon" that arises as a result of the complexity of neuronal connections. This theory sees the brain as a three pound computer made of meat. This is the dominant view held by a majority of scientists and is being pursued vigorously through the new technologies of Positron Emission Tomography scanning (PET) and the more effective technology of functional Magnetic Resonance Imaging (fMRI).

fMRI can correlate specific brain regions with virtually all neuronal processes. These include vision, speech, memory and all bodily inputs such as touch, pain hot and cold sensation, as well as emotional responses. The idea is to completely map out the neural correlates of all subjective sensations. They feel that by doing this, they can (in the words of Wikipedia) "explain the exact relationship between subjective mental states and brain states [and] the nature of the relationship between the conscious mind and the electro-chemical interactions in the body." In a way, it's like taking apart a complex clock in order to understand time. This assumption has two rather large caveats.

Firstly, even though taking apart a clock may explain how it keeps time, it still doesn't explain what time is. Even Einstein could do no better than to explain that time is what you measure with a clock. Today, quantum mechanics has shown that time doesn't even exist on the quantum level, and it may even run backwards. The same reasoning applies to consciousness. You can't dismantle a brain and find out how consciousness works. Even if you could create a robot that could mimic human consciousness perfectly, that doesn't mean that the robot is actually self-aware any more than a clock actually creates the time it keeps.

Secondly, according to quantum theory, the existence of matter itself relies FIRST on observation by a conscious entity. In other words, consciousness came first, and THEN matter, the stuff out of which a brain is built!

As we previously explained, prior to the beginning of the twentieth century, all sciences were built on a pyramid with a foundation based on empirical facts. For example, the science of psychology rests on facts built up on the science of biology, which itself was built up on the science of biochemistry which rests on the science of organic chemistry which finally rests on atomic physics (how atoms interact with each other). The physics itself rests upon classical physics, the solid earthly rock of everlasting particles with definite momentum and velocities.

After Einstein, Schrodinger, Von Neumann and Bohr, the everlasting particles were gone. The solid rock of materialism has turned into an unexplainable quantum miasma. Heisenberg delivered the fatal blow, and quantum physics has proven that objectively real subatomic particles pop into reality only when observed by a conscious entity. Consciousness first, material brains second!

Quantum and mathematical attempts to deal with consciousness

The disconnect between everyday macroscopic reality, which seems to exist without conscious observation, and quantum reality which doesn't, has spawned a number of different theories which see quantum effects as a way out of the disconnect between reality and the quantum world. This is actually an active area of research by a number of prominent theoretical physicists. The names to Google in the field of quantum consciousness are David Bohm, Roger Penrose, Stuart Hameroff and Max Tegmark. Their work involves the possible quantum structure of the parts of the brain and their component neurons.

Penrose and Hamerhoff

Roger Penrose and Stuart Hameroff argue that a physical brain can collapse a massive grouping of microscopic wave functions into a real macroscopic object *prior* to its coming to consciousness. A brain can do this, not because it is conscious, but because it joins into a superposition state with the object. This, they believe is the result of "quantum gravity effects in microtubules." Biologically, microtubules are tiny structures within all cells that act as a sort of skeletal transport system and help to maintain the cell's three dimensional structure. They are made from a protein called tubulin. Microtubules are also active in a multiplicity of other cellular processes including the physical construction of other cellular components.

Penrose and Hameroff propose a physical process *beyond quantum physics* that can collapse macroscopic superpositions *prior* to their being perceived. They believe that certain tubulin proteins within neurons can switch between two different geometries and therefore different *gravitational* states. Their theory suggests that this change in geometry can cause the new configuration to enter into a "long range superposition" with external objects thus collapsing the wave function. This proposal eliminates the need for consciousness to collapse a superposition state. It happens simply because a non-living protein collapses it *before* conscious perception. All you need (apparently) is tubulin in nerve cells. Note that this theory remains outside of the quantum paradigm because gravity is the one force that has not yet been integrated into a grand unified theory.

Aside from the fact that quantum gravity effects have not yet been observed, the concept of quantum gravity effects in microtubules has been attacked by many mainstream brain/consciousness researchers as being very unlikely. Max Tegmark wrote a well-received paper criticizing Penrose's approach as well as all approaches involving quantum effects pointing out that quantum effects are on the order of ten billion times faster than the time of neuron firing and excitation.

Even so, Penrose is a very well respected physicist. His 1989 book, The Emperor's New Mind argues that the brain is NOT like a computer in that it does not use algorithms to compute its output. This was a disappointment for most mainstream scientists who still insist that the brain is really just a complex computer. Penrose concluded that consciousness might instead be based on quantum randomness.

Max Tegmark

Max Tegmark is a theoretical physicist at the Massachusetts Institute of Technology who has formulated a mathematical model of consciousness based on quantum mechanics and information theory. He believes that consciousness is just another state of matter, like solids, gases and liquids. He proposes that a system demonstrating consciousness must have two specific traits; the ability to store and process huge amounts of information, and a system that stores information holistically. (In other words, it is so integrated that individual bits of information cannot be separated from each other.) He also proposes two new forms of matter; *computronium* which can process information much faster than our most recent computers (38 orders of magnitude faster to be exact) and *perceptronium,* a substance that "feels" self-aware.

The beauty of this particular system of consciousness is that it is the only one that tries to account for self-awareness, and the fact the entire thing can be mathematically modeled. The problem with it is that there are an infinite number of mathematical solutions that include all possible quantum mechanical outcomes and many other even more exotic possibilities. There are other small problems as well. Neither computronium nor perceptronium have ever been observed. It is also difficult to conceive of a state of matter like perceptronium that can "feel" anything.

The fact that this system cannot distinguish between a glass of water and an ice cube that floats in it should probably disqualify the theory from serious consideration, but the fact that it could explain consciousness without resorting to free will or a mysterious non-physical entity that can observe the universe is a tremendous plus to the community of physicists. None of them actually take it seriously, but it's a wonderful way to end arguments with an uninformed public.

Artificial intelligence

Finally, artificial intelligence is a field of study which skeptics like to think will finally explain (or at least replace) consciousness. The most salient thing that one can say about artificial intelligence is that it is artificial. Unless someone can come up with an actual substance like Max Tegmark's perceptonium, nothing in computing can actually be self-aware. No matter how realistic any program is, or how fast any computer can run, no one can escape the fact that artificial intelligence can only *mimic* a conscious entity. Your GPS can talk to you in a realistic voice, and with more complex programming and faster processors, it could conceivably answer your questions and sound as if it "cares." But at the end of your trip, you turn it off and it doesn't care a whit. It won't fall in love or abandon you for another driver. It's just a machine, and even the most advanced robot with an artificial body created with the look and feel of a beautiful man or woman and twenty quantum computer processors for a brain will still be as artificial as the voice in your GPS.

Furthermore, no matter how much information the computer has access to, it's "opinions" will always rely on the opinions of the programmers who write the algorithms. The most recent foray into artificial intelligence is Chat GPT, an online search engine that can hold conversations and answer complex questions in plain English (or any other language). My own experience with it has shown me that its "opinions" are those of the programmers, who suppress some sources of information while promoting others. The problem with a robot that uses the entire internet to form its answers to questions is that it has to be "selective" about its sources, but the selection process is in the hands of the programmers and whoever employs them. You have only one mind, but it shouldn't be controlled by a cabal of self appointed intellectual elites.

In the end, no matter how well you can simulate a human mind complete with simulated emotions using a computer program, it will still not be conscious. Christof Koch, a neuroscientist and the chief scientific officer at the Allen Institute for Neuroscience in Seattle put it this way: "You can simulate weather in a computer, but it will never be wet."

Why scientists are afraid to look for consciousness outside of matter

Every scientific alternative to the Copenhagen interpretation's demand for conscious observers is meant exclusively to allow physicists to avoid dealing with consciousness. Furthermore, physicists in general never bring up the subject in public because of the spiritual implications, although, it isn't so much the spiritual implications as the misuse of those implications that scares them.

For example, one movie, "What the bleep do we know?", offered a fair and graphic representation of the weirdness of the quantum world, but then veered off into spiritual revelation which was frankly outrageous, even to most fair minded and spiritually aware people. This isn't much different than the misuse of biblical quotes by unscrupulous nineteenth century preachers, or modern cult leaders.

Make no mistake! The implications of modern

quantum theory DO support an interpretation of reality that requires *not just* consciousness, but a consciousness external to the reality that can be studied by modern physics. However, science cannot delve into that mystery. The reason is that consciousness can not, itself, be mathematically modeled. There is no way that consciousness's centrality in quantum theory can be harnessed by engineers to create money-making devices. When consciousness poked its ugly head into the realm of quantum theory, academics and even the scientists who made the discoveries tried to disavow it.

On the other hand, academics and scientists of all stripes encouraged the study of cosmic inflation as soon as it was hypothesized because it can be mathematically modeled. In addition, even though no one has yet been able to apply cosmic inflation to any practical device, it brings in a ton of money in research grants.

Science is the discipline used to study natural phenomena that can be treated with tangible and testable models. Its findings must fit comfortably into the existing body of accepted scientific fact. Unfortunately, the implications of quantum theory regarding consciousness cannot be fitted into the existing body of accepted scientific fact. As we have seen above, consciousness is not easily defined. No one can pinpoint its source. There's no scientific test to prove it even exists. There's only the old philosophical proposition by René Descartes "I think, therefore I am".

The implications of quantum theory concerning conscious observers offend the sensibilities of academics who don't want to disturb the sacred tenants of the Copernican Principle.

6
The brain vs the mind

The Mapping of the Human Brain

In the early 1950's American-Canadian neurosurgeon Wilder G. Penfield (not to be confused with Roger Penrose discussed in chapter 5) launched an exploration into the neurological mapping of the human brain. Prior to this time, all brain mapping had been done using live animals and by correlating the effects of human brain injuries with abnormal function and behavior.

Because of advances in surgical procedures, and the knowledge that the brain itself has no pain receptors, modern surgical techniques could now be used to remove brain lesions that are the foci of epileptic seizures. These operations were done with local anesthetics and without putting the patient to sleep. Before the age of computers and fMRI, this had the advantage of allowing the patient to help guide the surgeon to the correct area of the brain.

With the patient awake, and the brain exposed, Penfield was able to use the electrical stimulation of tiny areas on the brain's surface to stimulate movements and sensations in various parts of the body. Besides the reactions he expected regarding muscular and sensory stimulation, Penfield was surprised that he could also elicit complex memories, speech and emotions, and he was able to map their locations as well. It seemed that these points were triggers for activating huge neuronal circuits. It was the first time anyone had been able to correlate higher brain functions and somatic sensations with specific areas on the brain's surface.

Penfield was able to stimulate visual, tactile and auditory hallucinations, as well as isolated elements often attributed to mystical experiences. His work was hailed for its scientific revelations correlating external objects, sounds and touch with the areas of the brain where we actually experience them.

However, these observations were used by people with less benign ambitions as proof that all spiritual sensations were generated by the brain and were not actually spiritual at all. Evidence like this appears to support the materialist view that spiritual experiences such as the appearance of ghosts, spirits, near-death experiences and even the sensations of telepathy and clairvoyance are also brain generated hallucinations.

But the ability to produce the vision of, say, a tree by electronically stimulating a brain does not mean that trees don't exist. It shouldn't be too surprising that Penfield found areas of our brain that are configured to allow us to see, feel or hear and recognize our external environment. Without them, we wouldn't be able to see, feel or hear! Neither is it too surprising that some brain areas are correlated with our ability to store complex memories. The ability to use electronic stimulation to reproduce them as hallucinatory phenomena doesn't mean that those memories never happened.

The ability to produce a hallucination does not mean that all similar sensations are hallucinations. Some artificial flowers are so realistic that it is difficult to tell them from the real thing. But the ability to fake a flower doesn't mean that all flowers are fake.

Likewise, the ability to artificially produce emotional elements of a near death experience does not mean that all near death experiences are hallucinations. In order to experience real spiritual phenomena, our brains HAVE to contain areas that allow us to experience them. Without the occipital areas at the back of our brains, we would be blind. Perhaps without these spiritually sensitive areas of our brains, we would be spiritually blind as well.

We'll be hearing a lot more about Wilder Penfield later in this chapter.

The Brain as Computer

The materialist view of the human brain is one of a physical machine designed by nature as a computer guidance system. It is essentially a central processing unit (CPU) with numerous peripheral accessory units which operate our skeletal muscles. It monitors and controls our bodies' smooth muscles and internal homeostatic systems in a largely unconscious way. It is hooked up to eyes (cameras) and ears (microphones) that allow us to gather external information. It integrates this information by using an "artificial intelligence" program commonly called consciousness. In other words, our consciousnesses (and what we call our souls) are just "epiphenomena", artifacts of the chaotic electrochemical processes of a computer made of meat. For all practical purposes, materialists view us simply as complex robots.

From a *spiritual* point of view, there's really no reason to doubt this interpretation. The brain really is a complex computer! It does almost everything that the materialists say it does. The real difference between the spiritualist and materialist perspective is not what the brain does for our bodies. It's where the programming comes from!

Over and above their raw electronic circuitry, real computers have two types of programming. First, there is the "Read-Only-Memory-Basic Input Output System"

(ROM BIOS). This is permanently embedded as an electrical component and oversees the internal regulation of the electronic signals. This includes functions like the interpretation of keyboard strokes and the formation of corresponding characters on a monitor along with numerous other automatic functions. It also tells the computer how to interpret the second type of programming which comes in the form of applications (apps) that we, as human beings use to perform the functions that make the computer useful to us. Apps are programs which are intentionally bought and downloaded by humans, and do not come with the computer.

Our brains have both types of programming. There's no question that the circuitry and ROM BIOS functions that the brain performs come with the human body. For example, we have circuitry that converts raw visual signals from our retinas into usable visual information that allows us to see. We also have ROM programming that allows us to integrate the signals into automatic responses that coordinate our muscles. For example, there is the cerebellum which integrates our proprioceptive senses and stimulates automatic muscle movements that allow us to keep our balance and coordinate our movements.

However the real usefulness of our brains (at least to humans in complex societies) comes from the second type of programming; the apps we intentionally download to make our brain-computers useful to us as humans. From the brain's point of view, we spend years downloading (learning about) our world. We, integrate the data, form opinions and finally conform our behaviors to match those opinions. This type of programming is NOT built-in. Two people subjected to exactly the same patterns of stimulation throughout their lives are likely to form completely different opinions about the world. No two people have exactly the same personality. The decisions they make throughout their lives lead to completely different outcomes. Science has no idea where this type of "programming" comes from, or why it is varies so much from person to person, but it is the only thing about our brains that makes us truly human.

Finally, we come to the question of intuition and psi (psychic insight, covered in the next chapter). Not only does science have no idea of where these uniquely human functions come from, but many scientists and atheists deny they even exist. Intuition is really communication with the human spirit. It's just one of the many forms of psychic insight. No one doubts that people have intuitive understandings that cannot be programmed into a computer. No computer has ever demonstrated psi or anything like intuition. People, however, have been gaining intuitive insights ever since they began walking on the earth.

*I believe in intuition and inspiration.
Imagination is more important than knowledge.
For knowledge is limited, whereas imagination
embraces the entire world, stimulating
progress, giving birth to evolution. Intuition is
the father of new knowledge, while empiricism
is nothing but an accumulation of old
knowledge. Intuition, not intellect, is the open
"sesame" of yourself."*

Albert Einstein

The point of all this is that the brain really <u>is</u> a computer. It comes with everything that makes a computer useful except the apps. It is the machine we are born with which allows us to take our first breath. The question is: "**Where do the apps come from?**"

The transmission theory of consciousness and the brain as radio

Materialists point to devices like computers and artificial intelligence as producing the equivalent of human intelligence. However computers or the GPS in your car only give the *illusion* of intelligence. No matter how complex the electronics or the programming, none of them are *self-aware*. They do not possess the property of consciousness.

On the other hand, WE are self-aware. And we are

conscious, even those of us who are mentally handicapped. Our conscious minds are not in the business of producing the illusion of intelligence. They really ARE intelligent.

But no one really knows where our self-awareness comes from. It hasn't been located in the brain, either by electrical stimulation or with any of the new digital imaging technologies. Maybe it doesn't exist inside the brain at all! After all, the long and credible history of communication with the dead through mediums, along with the well documented history of ghosts suggests that the conscious mind can exist in a spiritual realm totally detached from human bodies.

Knowing this, it's not that much of a stretch to believe that our own conscious minds exist in that realm as well. When it comes to consciousness, perhaps the brain acts more like a radio than a computer. The brain may simply be a radio tuned to channel our own souls. Real radios are not intelligent. The intelligence that spills forth from the speakers lies in a city far away, and only manifests in the radio. The same could be said for our conscious self-awareness.

The idea that the brain acts simply as a link between the physical body and a remote spiritual soul goes back to Hippocrates who argued that the brain is the messenger to consciousness. It carries on a two way conversation between the body and the soul. This is called *the transmission theory of consciousness.*

Since damage to various areas of the brain causes bizarre behaviors and altered consciousness, materialists like to argue that this is proof that the brain actually creates consciousness, and the thing that we call a soul is just a figment of our imaginations. Their assumption is that what we perceive as consciousness is created in the brain, and if the brain is damaged, then the source of consciousness is also damaged.

However, if you damage an electrical component in a radio, the voice of the announcer stops, or is distorted. It may even become a completely different voice. That, however, does not mean that the source of the voice has been damaged. The announcer remains untouched in his or her broadcasting booth miles away.

The transmission theory postulates that the same thing happens when an accident or disease that damages the brain alters the way a person perceives, thinks and acts. Defects in our brains may alter our ability to channel and act upon our spiritual instincts, but that does not alter the source of consciousness, any more than a damaged radio alters the mind of the radio announcer. Understanding this may be helpful to persons with loved ones facing dementia, Alzheimer's disease, debilitating strokes, or mental disorders that produce distorted or diminished personalities. People with these disorders are still "there." Their personalities still exist in all their vitality. It's just that their bodies are no longer capable of containing and channeling the full extent of their being.

For all the great strides in its knowledge of the anatomy, physiology and neural connectivity of the human brain, modern medicine still does not claim to understand how the human mind relates to the brain. Nor can it lay claim to the ability to cure mental illnesses. The symptoms of psychoses such as schizophrenia and affective disorders like bipolar disorder, OCD, and ADHD may be controlled up to a point using certain medications, but the underlying mental abnormality rages on even if the symptoms are masked. The relationship between mental illness and Spirit is fascinating and enlightening. This is examined in detail in book 2 of this series (The Structure of Heaven by Martin Spiller).

Wilder Penfield and the transmission theory of consciousness

Wilder Penfield was the Nobel Prize winning neurosurgeon we talked about earlier in this chapter. We did not mention at that point that Penfield, the pioneer who first mapped the brain came to the conclusion that the brain does not create consciousness, but rather acts as a transmitter for it.

As you recall, Penfield discovered that electrical

stimulation of certain areas on the surface of the brain could stimulate memories of events that happened years or even decades earlier. The patient would actually relive the event in his mind, in more detail than he or she could under normal circumstances. However, even while recalling the event in detail the patient would remain aware of everything that was happening in the operating room, and there was never any confusion between the two streams of consciousness.. (Remember that the brain has no pain sensors and Wilder performed his surgery on conscious patients so the patient could help guide the surgeon to the site which initiated the epileptic seizure.)

While Penfield could stimulate memory streams like this, as well as involuntary muscular movements, he was never able to affect the patient's awareness of his surroundings or any other function of the mind itself. He could produce tactile, visual and auditory hallucinations, but he could not produce alien beliefs, or affect decisions by electrical stimulation of any brain part. The closest he could come to affecting the mind itself was to turn it off entirely by producing unconsciousness when he stimulated the upper part of the brain stem.

After many years of trying to find out where consciousness lived in the human brain, _Penfield finally concluded that the mind only "interacts with" the brain at the level of the brain stem._ Ultimately, he decided that the brain itself was only a computer that could interface with both the mind and the body, and was not responsible for the mind itself.

> *"What the mind does is different [than what the brain does]. It is not to be accounted for by any neuronal mechanism that I can discover. There is no area of gray matter, as far as my experience goes, in which local epileptic discharge brings to pass what could be called "mindaction". There is no valid evidence that either epileptic discharge or electrical stimulation can activate the mind... None of the actions that we attribute to the mind has been initiated by electrode stimulation or epileptic discharge. If there were a mechanism in the brain that could do what the mind does, one might expect that the mechanism would betray its presence in a convincing manner by some better evidence of epileptic or electrode activation."*

Penfield, Wilder, <u>the Mystery of the Mind</u>,

What Is Reality?

So, maybe our brains do not produce consciousness, but only channel it. Maybe the Copenhagen Interpretation's most outrageous conclusion needs to be taken seriously. Perhaps the words of the Beetle's famous refrain, "Nothing is real" is literally true; unless, of course, it has been observed by a conscious entity. This tenured conclusion was something that even Einstein really didn't want to hear because it placed consciousness back at the center of the universe where it stood long before the rise of Newtonian physics. It interfered with the previously unassailable Copernican Principle (which moved human beings from the center of the universe to a position of less importance than a grain of sand).

In 1633 Galileo was summoned before the Inquisition to answer charges of heresy. He had turned a popular toy, the telescope, toward the heavens and discovered that Jupiter had moons. The church had written the book on popular seventeenth century political correctness. God created Man, Man was the center of the physical universe, and the church wasn't about to allow Galileo and his moons, or any other scientific upstart to upset their apple cart!

In 1925, quantum mechanics was invented in response to work by Albert Einstein, Werner Heisenberg, Max Born and Erwin Schrodinger. It implicitly stated that the reality experienced by all normal human beings was based upon a deeper reality which required conscious observation in order to materialize. Now a conscious entity was again at the center of the universe. Today, modern materialists and atheists have written the book on popular twenty-first century political correctness, and they aren't about to allow Erwin Schrodinger or any other scientific upstart to upset their apple cart.

But let's secretly ask the politically *incorrect* questions.

If the reality we experience is the result of our own conscious observations of the universe, did we create our universe through our own conscious observations? Or did the universe even exist before we were around to observe it? And if it existed before us, who's consciousness "collapsed the wave functions" and created the universe in the first place? You really don't have to be a scientist to take a stab at answering that question. Or perhaps you believe there really is a legion of Boltzmann brains floating around in outer space.

Most materialists, skeptics and atheists simply ignore the implications of quantum theory, partly because their entire worldview is based on easily understood but old fashioned classical physics. For the vast majority of materialists, the study of quantum reality requires an understanding of complex subject matter that they really don't want to spend the time understanding. If they think about the implications of quantum physics at all, they chock these mysteries off to philosophy or religion, even though they spring from the most potent physical science ever invented by the mind of man! But the question of who provided the consciousness to

create a universe in a determinate state remains
unanswered, and has never really gone away. It lurks in
the background of science like the ghost of
Schrodinger's cat.

Science no longer supports all the assumptions of
classical physics. No longer can materialists claim that
matter is composed of eternally existing physical
particles. They can no longer claim that if we could
know the position, mass and velocity of every particle in
the universe then the universe would hold no more
secrets and we would be as omniscient as God Himself.
Quantum physics has shown conclusively that we can
never simultaneously know the position and velocity of
any particle, or indeed, even if the particle exists at all
unless we look at it. In fact, Science now implies that
reality as we know it is a rather persistent illusion based
on conscious observation, and that consciousness is
much more fundamental than just an artifact of a
complex arrangement of neurons.

The skeptics and atheists have embraced the
multiverse theology as an alternative to a God created
universe, however, neither the theory of inflation nor
the multiverses that spring from it can boast any proof
that they are more than mere theologies. Nor can they
answer the question of why quantum physics places
consciousness squarely at the root of our reality.

I would never personally presume to have been able

to converse with Einstein, but I now can sympathize with the concerns he expressed during his meeting with Rosenblum. I have come to believe that the reality we experience every day is, in fact, an illusion created for us by "the old one"; that it is part of a much greater reality; that our world is only one illusion within a spiritual realm filled with illusory worlds.

7
Parapsychology

As the name implies, parapsychology is a subdivision of psychology. Parapsychology studies the subtle properties of the mind that are not objective enough to be studied by other branches of psychology. These properties involve things that are not quantifiable such as intuition, or the ability to sense information not available through any of the five senses.

In capsule form, parapsychology studies _four_ specific facilities of the human mind which are collectively called **psi processes** (pronounced "sigh"). These facilities are **telepathy, clairvoyance, precognition and psychokinesis**. All four are properties of living persons (and possibly some animals). While we have not, as yet broached the subject of the human soul, these human aptitudes are also the very facilities which allow a two directional link between the living and the dead. In this chapter, we will begin to discuss the transition between our reality and the next.

Note: This discussion of psi processes bleeds over into the spiritual realm simply because psi processes are the facilities that discarnates (the souls of the dead who now exist in the spiritual realm) use to communicate with each other and with living people, especially with those people called "psychics" and those with latent psychic talents. The world of Spirit, psychics and mediums is the subject of the second book in this series (The Structure of Heaven by Martin Spiller).

Telepathy

Telepathy is the ability of one human mind to communicate directly with another without visual, auditory or any other physical cues. Telepathy can convey information between two or more people over virtually any distance. Spiritualists believe that this facility also allows the mind-to-mind transfer of information from a discarnate spirit (a dead person without a physical body) to a living agent, and vice versa.

The Ganzfeld

Charles Honorton was a parapsychologist who worked as an assistant for J.B. Rhine at Duke University and later at the Maimonides Dream Lab. His work at the dream lab convinced him that psi effects have frequently been associated not only with dreaming but with other altered states of consciousness, such as meditation and hypnosis. Anecdotal reports and experimental findings led Honorton to postulate that psi may operate as a *weak* signal that is normally masked by the stronger signals constantly besieging us from our conventional sense organs.

The trick was to find a way to bring about a *hypnogogic* state in an experimental setting where it could be studied under controlled conditions. (A hypnogogic state is a state of consciousness between waking and sleeping. It is the semi-conscious state that one enters just before one is falling asleep. A *hypnopompic* state is the semi-conscious state that one enters just before wakening from sleep. Both are mental states in which the human mind is most susceptible to psychic influences.) Dream labs are incredibly expensive to run, so Honorton looked for an inexpensive way to produce a compatible altered mental state. He hit upon *sensory deprivation*. The idea was to find a way to reduce ordinary sensory noise while keeping the subject relaxed and receptive to the weak signals associated with psi.

In the ganzfeld procedure, a subject is placed into a comfortable reclining chair, each eye masked with one half of a ping-pong ball. A red lamp shines from above while "white noise" (soft electronic hissing) is played through earphones. After about fifteen minutes, a person in *another* sealed room (the "sender") opens a sealed envelope containing a target picture, or begins to watch a randomly chosen video clip. He or she then begins to concentrate on telepathically conveying images of the target picture or video to the relaxed subject in the other sealed room.

After the session ends, the subject is shown the target image or video, along with a set of three other randomly chosen pictures or video clips. The subject then chooses the one he or she thinks it is to the target that the sender was trying to convey.

We would expect the number of hits to be 25% if the choice was totally random. A study in which the number of hits is statistically greater than 25% would be considered a successful study. Many studies were carried out and an amazing 55% of the studies reported statistically significant results (greater than 25% hits), whereas only 5% would have been expected to do so if chance alone had been operating.

Note: The process of combining the statistics of numerous experiments is called "meta-analysis". This type of analysis is used in areas in which the signal being studied is weak and requires more subjects than can be accumulated for a single experiment. Psi facilities in an ordinary population are notoriously weak and difficult to study. Studies like this are expensive and time consuming, and budgets are generally very tight which is the reason that numerous small studies must be combined to get convincing results.

The skeptics, of course, rejected these findings, as

they always do. Honorton recalculated the results omitting certain studies to meet the skeptics' objections and still came out with an overall hit rate of 35%.

The skeptics continue to find new ways of rejecting results like these, mostly by constantly moving the goalposts, but also by simply refusing to look at the data. They reasoned that "Since psi phenomena are impossible, they can't happen. Therefore, something must be wrong with the results, or the researchers are simply lying."

Their ultimate objection is the "file drawer" fraud which hypothesizes that only successful studies are reported and unsuccessful ones are simply filed away and forgotten. Susan Blackmore, a famously skeptical parapsychologist, addressed this issue and could find no selective bias. Ultimately, she gave up trying to debunk parapsychological studies, but the rest of the skeptics wouldn't budge.

Anyone who is fair minded will recognize the results for what they are; proof that telepathy is a real, albeit weak human facility.

Clairvoyance

Clairvoyance is the ability of a human mind to gain information about a distant environment without direct visual, tactile or auditory cues. This facility is the basis of what is now known as "remote viewing." Technically, clairvoyance involves "seeing" at a distance, but the term is also a broad category that includes:

Clairaudience (hearing at a distance),
Clairsentience (feeling at a distance),
Claircognizance (knowing at a distance),
Clairgustance (tasting at a distance), and
Clairalience jokingly called clairsniffance", (smelling something without using the nose)

These are the facilities mediums and other psychic people cite as the way they see, hear or sense a discarnate spirit. All mediums have telepathic abilities as well as one or any mixture of the above "clairs." Some can visualize the spirit of a deceased person (clairvoyance), some can hear them (clairaudience), some simply sense their presence and gain information from them in a holistic way (claircognizance), and some can feel the emotions, the joy, or the pain, and discomfort of the spirits they channel (clairsentience).

Clairvoyance also works in reverse, and is the facility that allows the living to communicate with the spirit. Spirits do not have physical eyes or ears, and must rely upon their spiritual senses. They rely exclusively upon these senses to communicate with other spirits as well as with us. Their ability to communicate with the living depends upon their energy level. This is a subject that we will be studying later.

Clairvoyance and the US Government

Beginning in the mid 1970's, the US government learned that the Russians were conducting experiments using clairvoyant psychics in an attempt to spy on the US and allied governments. This was at the height of the cold war and not wanting the Russians to gain an advantage over the Americans, the CIA decided to do experiments of their own to counter the Russians. From 1978 to 1995, the US government ran its *Stargate* program studying *remote viewing* which was defined as "the ability of human participants to acquire information about spatially and temporally remote geographical targets, otherwise inaccessible by known sensory means." ("Temporal" means separated in time.) The results of this program were decidedly mixed. There were quite a few spectacular successes, but few of the successes were without at least a few failed components.

One psychic, sitting in California visualized details inside a secret National Security Agency listening post in West Virginia, right down to the words on file folders. Another psychic nailed the code name for a remote site and its physical layout but botched the names of people working at the site. That same remote viewer, provided with only map coordinates and an atlas, drew a map containing new buildings and a massive construction crane hidden at a secret Soviet nuclear weapons facility. The map he drew was subsequently proved to be uncannily accurate according to U2 spyplane imaging. Unfortunately, he got most other details wrong.

In 1981 Army Brigadier General James Dozier was kidnapped by the Red Brigades in Italy. Six hours after the kidnapping, remote viewers, including Joe McGoneagle were given a photo and his name and asked to describe his location. McGoneagle said that he was located in Padua Italy while another remote viewer gave the name of the building. In the meantime, Italian authorities had gotten information from a relative of one of the kidnappers and rescued Dozier. It was later reported that the information provided by the remote viewers was accurate, "right down to the bed where he had been chained."

In the end, the CIA decided that the information provided by remote viewers was too unreliable to use for practical intelligence gathering. By 1995, the lack of practical applications persuaded a number of skeptics in congress to drastically restrict the program's funding. The skeptics placed Ray Hyman, perhaps the most skeptical secular humanist in the world and one of the founding members of CSICOP in charge of the program which effectively ended the government's foray into psychic research.

Precognition

In 1898, a novella entitled "Futility" was published by

author Morgan Robertson. It was a fictional story about a gigantic ship, 800 feet long, and, clad with iron, considered to be unsinkable. Because it was thought to be unsinkable, it carried only twenty four lifeboats, the minimum required by law, even though the ship's capacity was 3000 passengers. It was built with nineteen watertight compartments separated by bulkheads that could automatically seal themselves in the event of an accident. In the novella, the fictional ship was sailing in the North Atlantic Ocean when just before midnight she struck an iceberg on her starboard side and sank with heavy loss of life as a result of the inadequate number of lifeboats.

Fourteen years later the very real RMS Titanic was launched by the British White Star Line. She was 882 feet long, clad with Iron and considered unsinkable. She was equipped with sixteen watertight compartments so that if one were flooded, the others would remain dry and prevent sinking. Despite a capacity of three thousand passengers, she was equipped with only twenty lifeboats. She was, after all, unsinkable. In the early morning hours of April 15, 1912, she collided with an iceberg on her starboard side and sank shortly afterward with heavy loss of life due to the inadequate number of lifeboats.

The most uncanny incident of all; the author of "Futility" had named his fictional ship the Titan.

This anecdote stretches the 'coincidence" explanation fairly thin. There are an awful lot of coincidences to explain. But flashes of the future are fairly common. The assassinations of both President John Kennedy and Senator Robert Kennedy were predicted by many people, most of whom did not even know they were psychic. President Reagan's attempted assassination was predicted, and even published thirteen days in advance by psychic Noreen Renier. Unfortunately, no one was able to convince authorities to take appropriate preventive action in any of these cases.

Precognition is the ability to see into the future without any currently available information that could lead a person to deduce that a future event might take place. As you learned in chapter 1, time doesn't even exist as a unidirectional property (at least on a quantum level) until it is observed by a conscious entity. Because we can't peer through the firewall between quantum reality and our own, it's really impossible to know all of the ramifications this might have. The only thing we DO know is that people get premonitions of real future events once in a while, and they can't all be attributed to coincidence.

Since the invention of computers, parapsychology has come a long way in studying this phenomenon using repeatable experiments and statistical analysis. In a paper written in 2011, Daryl J. Bem of Cornell University discusses some of these experiments:

> *Physiological indices of participants' emotional arousal were monitored as participants viewed a series of pictures on a computer screen. Most of the pictures were emotionally neutral, but a highly arousing negative or erotic image was displayed on randomly selected trials. As expected, strong emotional arousal occurred when these images appeared on the screen, but the remarkable finding is that the increased arousal was observed to occur a few seconds before the picture appeared, before the computer has even selected the picture to be displayed. The presentiment effect has also been demonstrated in an fMRI [functional Magnetic Resonance Imaging] experiment that monitored brain activity (Bierman & Scholte, 2002) and in experiments using bursts of noise rather than visual images as the arousing stimuli.*
> (Spottiswoode & May, 2003)

> *Across all 100 sessions, participants correctly identified the future position of the erotic pictures significantly more frequently than the 50% hit rate expected by chance: 53.1%, t(99) = 2.51, p = .01, d = 0.25.3.*

> Feeling the Future: Experimental Evidence for Anomalous Retroactive Influences on Cognition and Affect

Science, Math and God

Daryl J. Bem
Cornell University

A deviation of 3.1% (53.1% vs. 50%) may not seem very exciting, but it is a gigantic deviation from random chance. Consider, for example, the studies to determine if aspirin reduced the risk of heart attack. These studies were discontinued after six years because it was considered unethical to keep this treatment away from the control group who had been given a placebo. Taking aspirin was found to reduce the risk of heart attack by less than 1% (.8% to be exact). The effects found in this series of precognition studies is nearly 4 times larger than the effect of aspirin on cardiovascular health.

Far from forbidding precognition, the fundamental laws of physics are completely neutral with regard to the direction of time. After pondering the implications of his special theory of relativity, Einstein wrote to a friend that the distinction between past and present was only a "stubbornly persistent illusion". Physicist Gerald Feinberg had this to say about precognition:

> "Instead of forbidding precognition from happening, [accepted physical] theories typically have sufficient symmetry (between past and future) to suggest that phenomena akin to precognition should occur. . . . Indeed, phenomena involving a reversed time order of cause and effect are

*generally excluded from consideration on the
ground that they have not been observed, rather
than because the theory forbids them. This
exclusion itself introduces an element of
asymmetry into the physical theories, which some
physicists have felt was improper or required
further explanation. . . . Thus, if such phenomena
[precognition] indeed occur, no change in the
fundamental equations of physics would be needed
to describe them."*

Is the future fixed?

Louisa Rhine, who some might call the "mother of
parapsychology" probed this question. Out of 433 cases
she found 191 in which people had attempted to
prevent a foreseen event from taking place. In 31% of
the cases, the attempted intervention was unsuccessful,
usually because the psychic experience had not
provided enough information to allow the person who
had experienced it to prevent the occurrence. However,
in the remaining cases (69 percent) the experient was
able to take adequate steps to prevent the event, or at
least to avoid some of the undesirable consequences.

Here is an example. In Louisa's own words:

*It concerns a mother who dreamed that two hours
later a violent storm would loosen a heavy
chandelier to fall directly on her baby's head lying*

*in a crib below it; in the dream she saw her baby
killed dead. She awoke her husband who said it was
a silly dream and that she should go back to sleep. .
. . The weather was so calm the dream did appear
ridiculous and she could have gone back to sleep.
But she did not. She went and brought the baby
back to her own bed. Two hours later just at the
time she specified, a storm caused the heavy light
fixture to fall right on where the baby's head had
been—but the baby was not there to be killed by it.*

Rhine, L. E., "Frequency of Types of Experience
in Spontaneous Precognition."

Smaller predicted events like an automobile accident
or a medical emergency are not set in stone. These may
be avoided or modified by individuals who take action
to prevent them. Louisa Rhine suggests that one should
try to prevent the events predicted in such precognitive
warnings, but because efforts to prevent disasters often
fail for reasons entirely beyond anyone's control, one
should not feel guilty or responsible for the failure.

Larger predicted events that affect the fates of large
numbers of people such as natural disasters, or the
assassination of important people are more likely to
occur regardless of the source of the warning. Even in
Spirit, the extent of precognitive abilities is actually
fairly limited. Large scale events seem to echo
backwards in time and are more easily foreseen, but

more difficult to prevent. For smaller scale events, time can be a factor. Even spirits cannot predict events over more than a short timeline. The reason for this is that everyone involved in the predicted event can use their own free will to change the course of events, even though most of the time, they have no idea that their decisions and actions are changing the future.

Psychokinesis

Psychokinesis (PK) is the use of the mind to directly move an object or alter its motion. This involves moving objects without touching them or influencing them in any physical way. Some of these events are large-scale and can be frightening, like the movement of objects during an exorcism, or a poltergeist haunting, but psychokinetic activity may also involve small-scale activities like the ability to influence the fall of dice or something as subtle as changing a bit in a computer memory.

The intention, or the ability to *will* the movement of objects in the physical world using psychokinesis is available to both the living and the discarnate, however all the *energy* used to actually move objects in the real world must come either from the physical environment or from a living person (also called a living **agent** by parapsychologists).

Psychokinetic events seem to follow some psychics

around, causing unexpected sounds, slamming doors and other ghostly movements. This is probably because these psychics create so much energy that it is always available to passing entities who pick it up like a football and run with it. This type of phenomenon is responsible for the physical mediumship displayed by some psychics. It is also responsible for the physical phenomena seen during apparitional hauntings and poltergeist cases when a person who is unconsciously a "psychic sensitive" is present.

In the late nineteenth and early twentieth century, public displays of physical mediumship were fairly common. Unfortunately many physical psychic phenomena can be faked by skillful magicians and the extent of fraudulent mediums exposed during this period contributed to the decline in belief in psychics and mediums in general. As a result, most physical mediums today do not publicize these abilities or try to perform as magicians. Nevertheless, psychokinetic phenomena still exist and are responsible for the poltergeist-like activity seen in active hauntings.

D.D. Home

The most famous physical medium in history was Daniel Dunglas Home (1833-1886) whose phenomena were spectacular. Most have never been duplicated. He held his séances in full daylight or in brightly lit rooms in intimate settings where guests could closely inspect everything he did.

Home never had a home of his own and made a living traveling from place to place and living as a guest for short periods in the homes of various friends. He held his séances in literally hundreds of homes and no one ever saw him prepare a room in any way. His séances would include a dozen sitters and start off with the entire room shaking, as if a small earthquake was occurring.

Musical instruments would levitate and play in broad daylight without any visible hands operating the keys or valves. One of the favorites was a hand accordion. Home would hold one end while the other end would expand and contract the bellows all by itself while playing tunes that often brought tears to the eyes of the sitters. A heavy oak table around which all the sitters were seated might tilt, with objects on the table staying in one place instead of sliding off.

His most famous feat was his own levitation. In broad daylight, Home reportedly went into a trance and floated out a third floor window, then floated back in through another window. This was observed by a number of stunned witnesses including three irreproachable members of London's high society, Lord Adare, his cousin Captain Charles Wynne and the Master of Lindsay.

There is no documented instance of anyone catching him cheating, although contemporary rumors of cheating have been seized upon and enlarged by modern skeptics to slur him. It is difficult to see how he could have cheated since the sittings were carried out in broad daylight, there was never any pre-preparation of the room and virtually thousands of people witnessed his feats in full light.

Even in his day, he was roundly hated by the journalist and intellectual class who were unanimously against the worldwide Spiritualist fads that followed on the heels of the events at the Fox residence in 1849. (See my book The Structure of Heaven *by Martin Spiller.)* The worst attack against Homes was carried out by the famous poet Robert Browning, an avowed skeptic. He wrote the well-known poem "Mr. Sludge, the Medium" which was widely considered an attack on Home. Browning's wife forced him to attended one of Home's sittings. Browning never claimed in public to have caught Home cheating during this sitting, and he even admitted in private that deception was out of the question, but he steadfastly refused to believe his own eyes.

The reason for this attack was assumed by most contemporaries to be jealously. Browning's wife, the equally famous poet, Elizabeth Barret-Browning was so wholeheartedly enthusiastic about Home's mediumship that their disagreement on the subject became a minor public scandal.

Homes never took credit for his psychokinetic feats, attributing them to friendly spirits over whom he had no control. Indeed, at one difficult point in his life, his spirits told him that they would abandon him for a year, and during that year he was unable to produce any physical phenomena. His "powers" were restored a year to the day after his spirits had stated their intention to leave him.

The Law of Psychic Phenomena

Parapsychology is considered by its practitioners to be a science, and while parapsychology has done a great deal to prove that the human mind is indeed capable of these facilities, *it does not concern itself with the question of survival of the spirit after death.*

Many parapsychologists take the position that since all four psi processes are generally weak facilities of the normal human psyche, there is no need to invoke the spirits of the dead to explain any information or psychokinetic activity that may be encountered via this route. This is not to say that all parapsychologists today reject the concept of discarnate spirits, but a substantial number still do. It is much easier today for an academic parapsychologist to make a living as a skeptic than it is to declare a belief in the paranormal.

In fact, there is no way to prove that "ghostly

phenomena", or information gained through psi processes does not derive either from the subconscious mind of the observer, or from telepathically conveyed information from another living person's mind. This unfortunate (but true) axiom originated with a man named *Thomson Jay Hudson* who wrote a book entitled The Law of Psychic Phenomena in 1893.

Hudson noticed similarities between hypnotically induced trances and the trances of spiritualist mediums. He observed that hypnotic subjects often possess enlarged extrasensory facilities that they did not have when in a normal state. From these observations, he formulated the hypothesis that any contact with "spirits" was in fact contact with the medium's own subconscious. Anything else could be explained by subconscious telepathy and clairvoyance. In the case of physical mediums and poltergeists, psychokinesis on the part of the medium could account for moving objects without invoking spirits.

In other words, if one believes in the human attributes of psi phenomena, it is easy to be skeptical of the existence of a spiritual world on the basis of mediumistic communication alone. Some of the skeptical argument against a spiritual world results from this type of reasoning.

There are, however, several caveats to Thomson Jay Hudson's law of psychic phenomena. Firstly, remember that all mediums are psychics. In other words they can gain information from the minds of their sitters as well as from the spirits that have come to visit them. It turns out that most mediums can feel a qualitative difference between the information gained psychically from the brain of a sitter and a true message from a spirit. They say that gaining information psychically from a living person is colorless and without attached emotion, sort of like being read a shopping list.

Information channeled from a spirit, however, is anything but colorless. When a beloved grandmother speaks to her granddaughter, the medium can feel the emotions, and gain information about grandma's subtle quirks and hidden habits. These feelings and emotions are qualitatively different from those emanating from the living sitter.

Secondly, there are many cases of *veridical* (also called *evidential*) information that come through psychic insight that could not be gleaned from the minds of anyone present because no one present had any pre-existing knowledge of the subject. The adjective *veridical* applies to information that the percipient could not have known beforehand, but later is found to be true.

Separating Psi from Survival

Prior to 1930 the study of telepathy, clairvoyance, precognition and psychokinesis was bound up with research aimed at proving or disproving the survival of the human soul after bodily death. In the late 1800's and early 1900's, luminaries in the field such as Henry Sidgwick, Sir William Crooks, Sir Oliver Lodge and F.W. Myers collected anecdotes and did investigations of a multitude of psychics and mediums, endorsing some and discrediting many. They were quite aware of the objections raised by Thompson Jay Hudson and did their best to rule out simple ESP on the part of the medium. Unfortunately, the possibility that mediums were receiving their information from the minds of the sitters or investigators during séances rather than from discarnate spirits always hung over their heads like a dark cloud.

Starting in 1930 and continuing through 1980, the field of parapsychology was dominated by J.B. Rhine and his wife Dr. Louisa Rhine at the Duke University Parapsychology Lab. The Rhines wanted to put the study of the paranormal on a scientific footing in order to take it out of the realm of mysticism, and ensconce it firmly in the embrace of science. In order to do so, they decided to use only reproducible methods that could be carried out in the lab where conditions could be meticulously controlled.

They made the decision early-on that they would not attempt to do research on survival of the spirit after death because of the impossibility of scientifically validating the source of the information, and also because of their belief that a combination of telepathy, clairvoyance, precognition and psychokinesis could explain all supernatural phenomena. (See reference to Thomson Jay Hudson above.) These, they believed, were traits possessed by living mortals, and, if they could prove that this was the case, there was no need to invoke the spirits of the dead to explain them. They also were aware that psychical (pronounced "sigh-kik-al") research up to that point had been burdened with disastrous incidents associated with psychic hucksters and fraudulent mediums, and they wanted to avoid scandal at all costs.

The decision to discount the possibility of spiritual influences set the table for all future research carried out in parapsychology. Ever since Bohr and the Copenhagen interpretation, scientists had made the decision that science should stop at the threshold of our reality. The Rhines likewise decided that parapsychology should be concerned only with the measurable *phenomena* of psi, and its connection with the human brain. In order to account for the psychokinetic phenomena which were associated with violent hauntings and exorcisms, they invented the term **"poltergeist"**. For skeptical parapsychologists, "poltergeist phenomena" is a catchall term used to describe all physical phenomena that most people attribute to ghosts and demons. These are things like objects thrown across the room by unseen hands, loud disembodied noises, pools of water mysteriously materializing on the floor...etc.

The Rhines came to a conclusion which made them skeptical of even veridical information that came from mediums or ghostly apparitions. (Remember that the adjective "veridical" refers to information that a psychic would have no way of knowing by normal means, but later proves to be true and accurate.)

Before the Rhines' research, telepathy was seen as a process that involved two people; a passive receiver and an active sender. In other words, the sender needed to actively transmit information so that the receiver could process it. The Rhines came to believe that this process was exactly reversed, with the receiver being the agent that actively dug information out of a passive sender's mind, or gained information directly from a remote environment. They based this assumption on letters from people who thought that the Rhines could help them with their own paranormal experiences. Many of the letters reported cases of clairvoyance in which there was no living active sender.

They concluded correctly that people had telepathic and clairvoyant experiences regardless of whether an active sender was involved or not. Therefore, they reasoned, when the spirit of a dead person appeared and imparted information about an environment or the circumstances in a living person's life, it was only a "hallucination" created by the receiver's subconscious mind in an effort to "rationalize" the source of the information. The information itself, they thought, would be based on evidence that the receiver had subconsciously dredged up through the dual processes of telepathy and clairvoyance.

Under this assumption, a medium's surprisingly accurate information about a sitter's dead husband was really the result of telepathically snooping through the sitter's mind, rather than information imparted through the husband's spirit. Additional information about any aspect of a remote environment or circumstances the sitter was unaware of could be obtained through a combination of clairvoyantly and telepathically rummaging through the minds of remote personalities, as well as the drawers in their desks.

This use of psi processes has been dubbed "**Super ESP**" by critics because it seems to push the otherwise weak psi facilities that living people were eventually proved to possess well beyond their capabilities. While there is little doubt that "remote viewing" allows clairvoyants to scout out territories unknown to them, the Rhines' research ultimately demonstrated that human clairvoyant facilities, though very real, were relatively weak.

Later research at Stanford University, and several years of experimentation conducted by the US government showed that human extrasensory perception is a real human facility but of little practical value for intentionally "snooping" or spying on enemy facilities. They found that human psi abilities are too weak and unreliable to yield information that did not require verification through other more common methods. This is the reason that the US government eventually abandoned psi research.

On the other hand, the easy, consistent and often spectacular results obtained by psychics and mediums during thousands of normal sittings suggests that there is more going on than just the human facility of "remote viewing." The difference appears to be the mediums' acknowledged use of spirits to convey information as opposed to the parapsychologists' assumption that the information was gleaned through human psi facilities.

Mediums are sometimes able to impart information unknown to any living person, or information extant only in arcane historical documents. They have been helpful in locating dead bodies and in helping to solve crimes. In some instances, the information imparted would be known only to the victim or to the perpetrator of the crime. In many cases, the only explanation for the psychic insights that made it possible to find a dead body would be communication from the victim him or herself. Spirits are, after all, active agents with memories of their own, and (as you will learn in the second book in this series) the capacity to travel instantaneously to any location just by thinking about being there. Spirit mediated information transfer offers a more facile mechanism of information exchange than human ESP alone.

And yet "Super ESP" has become a common explanation for spiritual phenomena in parapsychological circles today. Thus, when a person suddenly sees the apparition of Uncle Harry saying goodbye at the same time he was dying a thousand miles away, many parapsychologists assume that it is simply a hallucination cooked up by the recipient's subconscious mind with the help of clairvoyance and telepathy.

True or not, the invention of Super ESP by parapsychologists was enough to trivialize everything that a medium might discern about a sitter or her dead relatives. This would include information neither the medium nor the sitter had any way of knowing. It also made it possible to dismiss even the most startlingly evidential information obtained from dreams and apparitions. After all, the information had to exist *somewhere*, otherwise it could never be verified. Therefore, a skeptical parapsychologist could assume that the psychic used super ESP to seek out the source, no matter how hidden it might be.

Why do so many scientists reject parapsychology?

Having ruled out the use of mediums or any of the other more "exciting" sources of paranormal information, the Rhines restricted themselves mostly to throwing dice, using Zenner cards (a deck of cards with stars, circles, boxes, crosses and wavy lines), random number generators and statistical analysis of the results..

Using these methods, the Rhines were successful in proving that telepathy, psychokinesis, clairvoyance and precognition were real phenomena, even though the statistical signals were weak and often required amplification through meta-analysis. (Meta-analysis is the use of statistical methods to combine the results of many similar studies into a single super-study.)

Weak or not, the Rhines unquestionably proved that paranormal facilities are real, if generally weak human attributes. The publication of his seminal work, <u>Extra-Sensory Perception after Sixty Years</u> in 1940 convinced most natural scientists of the reality of psi phenomena, and Rhine's work on psi has since been confirmed by virtually thousands of studies using more modern techniques.

But the public really wanted to explore the question of survival, and this was something the Rhines refused to do. The Rhines only wanted scientifically reproducible results. Anecdotal evidence, no matter how dramatic, is not reproducible. Eventually, the public began to lose interest in their work. The Rhines were not talking to dead people, studying poltergeists, or investigating cases in which a high school girl kills the kids in her school by setting the gym on fire with her psychic powers!

Furthermore, the Skeptics constantly challenged results and, after researchers had met their demands, they moved the goalposts by demanding even more stringent conditions. The skeptics never did any research of their own. They were always content to sit on the sidelines hurling insults and demanding conditions for experiments that were beyond the means of even scientists studying the physical sciences.

Statistics analyzing the results of Zenner card trials and dice throws make for dull reading and poor story lines for movies. Over the years, as the public interest in parapsychology flagged, the skeptics became bolder. This resulted in the slow elimination of the lab's funding and a boycott by influential members of the scientific community. The Rhines were forced to leave the campus of Duke University and move across the street into private quarters to continue their research.

This result has much less to say about psychic phenomena than about the political success of closed minded skeptics who are mired in eighteenth century Humean materialism.

The Cross Correspondences

One experiment in particular went a long way toward disproving the legitimacy of the theory of "super ESP", as well as substantiating the legitimacy of spiritual survival. Until this chapter, I have avoided delving into the subject of the spiritual realm. However the subject of parapsychology walks the dividing line between our everyday reality, and the world of spirit. The purpose of this book is not to prove the existence of a spiritual realm, but, rather to allow you to walk up to the boundary of our reality and peer over the edge. You can't believe in God without first understanding that He, as well as all of His subordinate spiritual entities have a place to live.

The protagonists of the following tale were very prominent and respected intellectuals who lived during the second half of the nineteenth century. They were also parapsychologists, (although the term was not invented until 1930). They spent most of their adult lives studying psi and the evidence for survival of the human spirit after the death of the body. They encountered the same problems as modern parapsychologists since the elites of their day were just as skeptical as modern elites when it came to ghosts, mediums, psychics, near-death experiences, and the mystical experiences of ordinary people. It appears that after a group of them had died, their spirits got together and fashioned an experiment that they believed would prove their continued spiritual existence, while at the same time, ruling out the skeptical arguments of super ESP and fraud. This is the story of that experiment.

Between 1901 and 1932, a series of spirit communications involving multiple spirits were transmitted to multiple mediums living and operating on different continents thousands of miles apart. The mediums involved were not familiar with one another and had no reason to communicate. By themselves, the individual spirit communications made little sense, and had no meaning to any of the mediums who received them. However, when fitted together by a panel of classical scholars, these otherwise meaningless scraps created coherent and intricate messages. This series of communications, known as the "Cross Correspondences" appears to have been designed by the spirits themselves to prove that they survived death.

The process of fitting the apparently random fragments together was similar to the assembly of a modern, compressed digital image received over the internet. The image you see on your computer monitor is composed of millions of bits of information which are sent out in "packets" of bytes. These packets are received by your computer in no particular order and any given packet means nothing by itself. It's the computer's job to unscramble, interpret and finally reassemble the packets in the correct order so that they can be presented on the monitor as a coherent image.

To make matters even more complex, the information conveyed in the cross correspondences was, in a way, encrypted. The ultimate messages were based on allusions to obscure classical literature that very few people other than classical scholars would have recognized. The use of obscure classical literature would have been consistent with the personalities of the spirits who apparently devised the concept of the cross correspondences, particularly Frederick W.H. Myers who died in 1901. His spirit appears to have been the instigator of this experiment. The reassembled messages used subtle symbolism, and each one might be dictated over a period of weeks, months or even years (thirty years in one instance).

These cases are rarely studied today because of the esoteric nature of the material transmitted and the incredible complexity of piecing the bits together. (Esoteric means "material that is intended for and understood by only an initiated few." These initiated few were the vanishing coterie of classical scholars.) Trying to describe even the simplest ones is so tedious that most moderns who are not familiar with classical literature have trouble following the logic. Nevertheless, the cross correspondences were a remarkable spiritual accomplishment and represent one of the most comprehensive proofs of spiritual survival ever produced.

Skeptics generally avoid talking about the cross correspondences, and feel justified in doing so because so few people today can understand the classical allusions and most can't follow the logic. However, when the skeptics are confronted by people who have studied and understand the evidence, they fall back on claims of fraud, collusion or "Super ESP." In addition, the study of the paranormal and spiritual survival today is very nearly universally reviled by the very academics that would ordinarily be called upon to study them. Atheism has become the standard among academics today, and they defend it as zealously as the fifteenth century Catholic Church defended catholic doctrine.

Where Parapsychology stands today

It's safe to say that the success of scientific parapsychology studies has proven beyond doubt that psi phenomena are real. In 1940, J.B. Rhine published a seminal work titled Extra-Sensory Perception after Sixty Years. It is still widely regarded as an influential work that grounded parapsychology firmly in science. Since the discovery of quantum physics, many scientists had begun to suspect that human psychic abilities might actually be real, but they were afraid to voice their beliefs. With the publication of Rhine's book, they felt vindicated and began to speak publicly in support of psi.

The dual theories of relativity and quantum mechanics, having gained universal acceptance, have created a new scientific paradigm. The primacy of consciousness and the discovery of non-locality have provided a framework that virtually demands the inclusion of paranormal phenomena in the body of mainstream science.

The *current* laws of physics are not violated by paranormal phenomena. Psi may not be compatible with the *old* scientific worldview based on Newtonian classical physics, but it is *perfectly compatible* with emerging worldview based on quantum mechanics.

In addition, the lack of public interest in the statistics necessary to scientifically evaluate the evidence uncovered by parapsychology does not extend to the paranormal phenomena themselves. Public interest in ghosts, poltergeists, psychics and mediums is as strong as ever! No matter how vigorously the elites try to stamp out belief in these very human facilities, a majority of people quietly, and often secretly accept them. Surveys have shown that over half of the adult population claims to have had psychic experiences and believes in the reality of the phenomena. In addition, whenever ordinary people hear about or experience an encounter with ghosts, poltergeists or a comforting encounter with a deceased loved one, it simply reinforces their native belief in in a world of spirit and by extension, in God. Who among us *really* wants to believe that "life sucks, and then you die"?

Studies have shown that psi facilities are real, but weak attributes of human beings. It may be that the apparent weakness of psychic facilities is due more to the cold, methodical methods used to study them. When they are observed in less formal settings they are much more impressive. Human paranormal facilities do not appear weak when psychics or mediums are working one-on-one with warm human clients or cold, dark ghosts. An emotional connection seems to be the key lacking in the scientific method. Recent work by Gary E. Schwartz, Ph.D. has produced some very impressive statistical evidence which turned out to be anything but weak. He tested a number of famous mediums using human sitters under strict test conditions. The results were impressive and were published in his book: The Afterlife Experiments.

8.

Can Quantum Theories Explain ESP?

Paranormal facilities like telepathy, clairvoyance, psychokinesis and precognition not only exist, but are essential human facilities. Parapsychology, although having gained a grudging acceptance, is still considered a sort of illegitimate step child of science, and most mainstream scientists still refuse to talk about it for fear of losing their funding. However, the fact that these very human faculties actually exist becomes evident once you study the literature of the paranormal and begin to understand the relationship of the world of matter to the world of spirit.

Every civilization in history was originally built around a moral code based upon a foundational religion, and every religion was built upon the revelations of prophets and sages who communicated with the dead. If you believe in God and in any of the metaphysical religions that claim to speak for him, you must also believe in the existence the four psi facilities. They provide the *link* between living human beings and the spiritual world. Clairvoyance and telepathy are the pathways through which they do this. Without these psi facilities there would be no way of learning about God, or spirits. Ghosts would have no way of manifesting, and there would be no history of people ever having seen one. People see ghosts because of clairvoyance, they hear them because of psychokinetic effects and clairaudience.

Non locality and psi

Once you understand the ramifications of Bells Theorem, it isn't much of a leap to conclude that the newly defined concept of non-locality and other aspects of quantum reality might somehow be involved in the transfer of information via telepathy and clairvoyance. At first, it may seem that this is the case, but it turns out, that it is no easier to relate paranormal phenomena to quantum reality than it is to relate it to our everyday reality. You can try, but you will get lost in a jungle of complexity and contradictions. I'll try to give you some idea of why this is true:

Quantum non-locality implies that particles across vast distances can communicate instantaneously without sending any signal across space. Experiments have shown that telepathy, like the nonlocal communication between entangled quantum particles, is indifferent to either the distance or the time difference between the sender and the receiver. Perhaps quantum non-locality can account for, or at least figure into telepathic communication between two individuals.

As we argued above, if two entangled particles can

transfer information instantaneously no matter how far apart they are, then perhaps the space between these particles does not actually exist. If this is the case for "entangled" particles, there is no quantum mechanical reason that it should not also be true of all particles.

Perhaps space really *is* an illusion and everything in the universe is in contact with everything else. Aunt Martha lives in Poughkeepsie but just received a premonition that something is wrong with her beloved niece who is visiting Paris. This is a fairly common phenomenon. It happens to a lot of people, but it seems to happen to Aunt Martha more than others in the family. Why does this type of thing happen?

Maybe Aunt Martha's telepathic link with her sick niece in Paris is due to their "quantum physical proximity". Instead of *receiving* messages *transmitted* by her niece, maybe Aunti and her niece are just feeling the *same* feelings and thinking the *same* thoughts because they actually occupy the same space. Farfetched??? Yes! BUT then everything about quantum physics is farfetched.

This, of course, begs the question; "If space doesn't really exist and everyone is in contact with everyone else, then why don't I feel the same things and think the same thoughts at the same time as everyone else in the universe?" If space as we know it is just an illusion, then theoretically, Aunt Martha should be able to think my thoughts and feel my pain. How about the drunk lying in the doorway across the alley? This doesn't appear to be the case (unless Aunt Martha is hiding something she doesn't want me to know about).

Maybe the zeroing out of space only applies to specific people who happen to be "entangled", but not to other people. After all, Auntie and her niece are related and Auntie used to babysit for her sister's children. In this case, Auntie and niece are entangled and can think the same thoughts, but Auntie and I are not. Then, the empathy would extend only between these two, but not to me. But then we have to ask, "How can psychics and mediums who have never met their clients before gain factual information about them?" If psychic abilities apply only to persons who happen to have previous quantum entanglements, why should telepathy and clairvoyance be operative for clients with whom the psychic is neither related nor previously involved?

The take-home from all this speculation is that if quantum non-locality is involved in the remote transfer of information from one brain to another, or from a remote environment to a brain, the mechanism by which it happens can only be associated with some extremely arcane properties of human consciousness. The intelligence involved in creating such a mechanism would have to exceed the intelligence of anyone occupying a human body.

Science vs. Spirit

We can observe quantum properties and we can speculate about their implications, but science has no way of exploring beyond the observations. There's a *firewall* between us and the meaning behind quantum reality.

None of this, of course, means that science won't eventually find a way to show that quantum non-locality is involved in the transfer of psychic information. If that were to happen, it would finally provide a mechanism of action for psi phenomena. Psi would finally be accepted by the scientific establishment because it would now be seen as a natural phenomenon. Parapsychology would get the recognition it deserves because it would become possible to attach the phenomena it studies to the rest of the natural sciences.

This triumph would not, however, affect the *spiritual* aspects of psychic phenomena. The scientific acceptance of psi would have no scientific bearing on questions concerning the existence of the spirits of the dead or the world they inhabit. These things lie outside of our own reality, beyond the firewall. Psi phenomena are just the means by which we communicate with spirits. Proving the reality of psi would not simultaneously prove the existence of the spirits themselves.

Non-locality vs materialism

Materialists, on the other hand, would find themselves in something of a bind if quantum non-locality was proven to be behind psi facilities. And even if the existence of human psi facilities is never proven, the non-locality predicted by quantum physics shines a severe light on the rather uncertain nature of reality itself. Quantum non-locality is one of the most startling findings of theoretical and experimental physics and it shakes the foundations of materialism down to its very foundation. Materialism, after all, requires a completely comprehensible reality in which space, as well as matter behaves in predictable ways.

The early discoveries of quantum physics, which eventually entirely displaced Newton's classical physics smashed the clockwork universe to smithereens. Matter is made of particles that don't even exist unless they are observed, and modern experimental evidence has now undermined even the concept of locality. Quantum space appears to be an illusion and exists only when someone looks to see if there is matter suspended in it. If it were eventually proven that macroscopic space is also an illusion, then materialists would have an even more difficult time proving that macroscopic matter is not itself also an illusion; a rather precarious intellectual position for the materialists.

Spirituality, however, does not place any limitations

on our definition of material reality. Those who understand Spirit as the ultimate reality see earthly reality as part of a larger continuum. The world of matter is only one of *many* realities, *all* of which are illusions. Only when taken together do these illusory worlds add up to **Reality**!

This may seem like a rather facile statement based upon a lack of scientific evidence, but is there really any difference between the lack of scientific data proving the reality of the spiritual world and the lack any mechanism tying quantum reality to the reality that we observe every day? (See the story of Bell's Theorem in chapter 2.)   Frankly, the findings of quantum mechanics supports the existence of a living spiritual world more firmly than the materialist's reliance on a clockwork universe supports a cold, dead material reality.

The spiritual interpretation of reality simply takes for granted that consciousness is the primary property of the universe and states unequivocally that reality, no matter what underlies it, has a purpose! It also has one further distinct advantage over scientific explanations. It requires no math.

The evidence for a spiritual world is really quite extensive. The purpose of *this* book is to present the reader with an understanding of the limitations inherent in the scientific/materialist philosophy that underlies our modern world, and to excavate a mental landscape that can allow an otherwise materialist mind to accept a spiritual reality. The evidence and rational for a spiritual world can be found in this book's sister volume: "The Structure of Heaven *by Martin Spiller.*

The next, and final chapter of this book will show you just how far modern skeptics will go to discourage your belief in spiritual values, and just how ludicrous their arguments are.

9
Skepticism

Humanism

Science took a great leap forward between the birth
of Galileo in 1564 and the death of Isaac Newton in
1727. Newton's classical physics paints a picture of the
universe as a gigantic clockwork mechanism operating
as a self-regulating machine, each part interacting with
all the other parts in a never-ending dance. (Please see
chapter 1 to refresh your memory.)

The *"**Enlightenmen**t"* was the scientific revolution's
intellectual counterpart. The most striking feature of
the Enlightenment was its rejection of dogma and
tradition in favor of the rule of reason in human affairs.

People in the eighteenth century (the 1700's)
remembered the horrors of the religious wars and the
Inquisition. The elites were especially ready for a
worldview that didn't include popes, bishops, bibles or
God. They found what they were looking for in the
philosophies of Diderot and Voltaire who argued for a
worldview of uncompromising materialism. The
doctrine of materialism takes classical physics as a
complete description of all of nature. It assumes that all
events are caused by *"causally-connected"* interactions
between particles of matter. The term "causally-

connected" implied that everything, including the human brain is made of particles, like billiard balls on a pool table. These particles constantly bounce around the table in a never ending mechanical dance. The billiard balls "connect" with each other "causing" an infinite series of movements that goes on forever.

Materialism views the human brain as just a mass of particles that happens to produce consciousness as an "epiphenomenon" which is defined as a sort of "unintended byproduct" produced by a hugely complex mass of neurons designed primarily to control bodily movement in response to environmental stressors. For materialists, free will is just an illusion because the clockwork universe was set in motion long before the earth formed. The destiny of all the atomic billiard balls was set at the beginning of time and can't be altered, and thus, everything you are, everything you do, everything you think and everything you will ever think was all pre-determined when the universe began.

Enter **David Hume** (1711 – 1776). Hume was an early eighteenth century Scottish philosopher and contemporary of Diderot who caught the ball passed to him by Voltaire and ran with it. He was the father of the philosophy known as *Humanism*. "**Secular Humanism**" is its primary descendant. Hume denied that humans have an actual conception of the self. He believed that what we think of as "experience" is only a bundle of sensations, and that the self is just a bundle of *"causally-connected* perceptions". He saw people simply as complex machines and consciousness as a product of the brain. Hume was an avowed atheist.

> *"The life of a man is of no greater importance to the universe than that of an oyster"*

Hume believed that the mind is simply a byproduct of the mechanical dance of particles in the brain. Therefore, every action, every thought, every decision, every kindness and every cruelty is simply the result of the pre-determined motions of the billiard balls. Everything is fated to happen, and nothing you can do can prevent it. Humanism sprang from these roots. According to Humanism, **free will** is a myth. All the decisions you will ever make in your life were pre-determined long before you were born. This is a very important point because, as you will see, if you believe **in Spirit,** *free will* **is EVERYTHING!**

CSICOP

The Committee for the scientific investigation of Claims of the Paranormal (CSICOP) (pronounced Psi-Cop but today known as the Committee for Skeptical Inquiry (CSI)) was formed in 1976 at a meeting of the American Humanist Association. The growing interest in all things paranormal alarmed Paul Kurtz, the editor of *The Humanist* magazine, and it was his idea to form CSICOP.

CSICOP is the modern enforcer of David Hume's philosophy. It is a *secular humanist* organization that masquerades as a scientific organization. It got off to a very good start because it pledged to be an *impartial* referee for scientific literature pertaining to the paranormal. It was *supposed* to be impartial, but many of its members knew that its real purpose was to put an end to all this nonsensical belief in spirits, God, and ultimately to religion, which they believed, in their hearts, was simply the lingering remnants of a bunch of superstitious nonsense left over from the dark ages.

CSICOP attracted a lot of famous academics including astronomer Carl Sagan, behaviorist psychologist B.F. Skinner, atheist philosopher Anthony Flew, author Martin Gardner, Harvard philosopher W.V. Quine and Marcello Truzzi, publisher of The Zetetic, a newsletter that dealt with academic research into anomalies and the paranormal. Truzzi's magazine was supposed to become the official magazine of CSICOP. The new organization also attracted a large number of stage magicians, including James Randi, perhaps the most skeptical stage magician in the world.

> *Note: it's a myth that a majority of stage magicians don't believe that people possess psi facilities. A number of recent polls have consistently found that between 72% and 87% of the magicians belonging to professional stage magician organizations believe in ESP. James Randi*

is more the exception than the rule.

    Over the course of the first year, it became apparent that CSICOP's central committee was dominated by hardline anti-paranormal crusaders who acted as grand inquisitors and who saw their job as stamping out any belief in the paranormal, and in essence, to promote atheism. This approach was in direct opposition to the *stated* goals of the organization which was to be an *impartial* referee. Once it became apparent that the organization was anything but impartial, a few of their prominent scientific members quit, including Truzzi who took his magazine with him. The organization then began to publish articles in *The Humanist* magazine, a publication of the older American Humanist Association. Later, CSICOP began its own journal which was appropriately named *The Skeptical Inquirer*.

    CSICOP's first, and only scientific investigation concerned the work of Michel and Françoise Gauquelin, a French husband and wife team whose main occupation had been debunking traditional astrology. The Gauquelins did, however, uncover some compelling statistics in support of one aspect of astrology; namely that the position of the planets at the time of birth actually does appear to correlate with certain human characteristics. Their most compelling positive result was called the "Mars Effect".

    The Gauquelins' results, published in 1975, just

before the formation of CSICOP showed that out of a population of 2,088 European Sports champions, 22% were born with Mars either rising or transiting. The probability of Mars being in any two specific sectors of the sky at one's time of birth is only 17%. The odds are millions to one against the 22% result occurring by chance.

Although the Gauquelins were mostly in the business of debunking the claims of traditional astrology, being good scientists, they did not withhold these results. The first attack against this claim came from an article in *The Humanist* magazine which objected to the Gauquelins scientific protocol and inferring that the statistics the Gauquelins had used were biased. The Gauquelins responded proving themselves to be more skilled in statistics than the author of the *Humanist* article, and in addition threatened a lawsuit against the magazine. Kurtz, the editor of *The Humanist* is said to have become frantic to attack the Mars Effect in print.

Other members of CSICOP jumped into the fray and suggested that the effect was probably due to an individual's time of birth since most births happen in the hours before dawn, and Mars appears near the sun more often than not. They proposed that the Gauquelins redo the study with NON-champions born at the same time of day as the champions to see if their birth rates also corresponded with Mars rising or in transit. If 22% of *non-champions* were also born at these astrological times, then the Gauquelins' finding on athletes would not be significant.

The Gauquelins repeated the study, this time for non-champions and controlling for location and time of birth. They duly delivered the results, but Kurtz withheld them for two years. When the results finally came out, they showed that the birth rate for <u>non</u>-champions during the times Mars was rising or in transit was 17%, exactly what would be expected by chance alone, and not the 22% that Kurtz and other CSICOP members had expected. However a second article in the same issue of *The Humanist* argued that the results weren't really valid because when female champions were dropped from the sample, the statistical significance of the results was reduced. Besides, most of the non-champions chosen for the second Gauquelin study were born in or around Paris, which for some reason further invalidated the results.

Two years later, the *Skeptical Inquirer* published the results of its own study on American sports champions which unequivocally *disproved* the Mars effect. But...

Shortly after the publication of the CSICOP study, astronomer Dennis Rawlins, who had earlier been excommunicated from CSICOP published an article called "Starbaby" in *Fate* magazine. Prior to his excommunication, he had been the only resident astronomer involved in the CSICOP Mars study.

According to Rawlins, when the CSICOP study had been completed and the original results came in, they supported the Gauquelins' study showing that 22% of *American* sports champions were born when Mars was either rising or in transit. However, Kurtz and others in the committee *manipulated the results,* distorting the data and making it appear that the study did not support the Mars Effect. Rawlins was thrown out of CSICOP because he had refused to go along with the ruse.

Rawlins final words on the subject were "I am still skeptical of the occult beliefs CSICOP was created to debunk. But I have changed my mind about the integrity of some of those who make a career of opposing occultism".

A number of independent investigations backed up Rawlins' charge, and eventually CSICOP admitted that it had "made mistakes". However they never acknowledged Rawlins's more serious charges that a Watergate style cover-up had occurred.

Several members resigned after reading Rawlins' article, and *CSICOP resolved never again to do another scientific study*. Therefore, today, most of the opposition faced by parapsychologists and other proponents of "unusual claims" comes from an organization that has been exposed cheating and as a result, now refuses to conduct any scientific research into the paranormal itself. Its refusal to actually DO science allows CSICOP to sit on the sidelines and hurl non-scientific accusations against scientific studies without actually having to defend its conclusions against charges of fraudulent science.

In 2006, CSICOP changed its name from Committee for the Scientific Investigation of Claims of the Paranormal to CSI, the "Committee for Scientific Inquiry" as a quiet acknowledgement that trying to use science to debunk the paranormal hasn't worked out well for them. It still doesn't perform any research of its own, and it has become simply a vigilante organization promoting a narrow brand of scientific fundamentalism. Its main occupation is to influence the media and public opinion. The name change allows it to masquerade as a non-biased scientific institute, but in reality it is really just a secular humanist organization with a huge budget. CSI, more than any other organization accounts for the public suppression of belief in the paranormal and the media's refusal to acknowledge religious belief as a necessary social institution.

Today, the skeptics attack all scientific studies concerning the paranormal as either flawed or fraudulent even if the studies and their statistical analysis are immaculate. CSICOP still doesn't do any scientific studies of its own in order to avoid scandals like "Starbaby". They stick pretty much to collecting money from their membership and vandalizing

Wikipedia and other popular media venues. They also send biased fools like James Randi around to various universities and colleges to spread the holy doctrines of secular Humanism and skepticism.

What's wrong with the Skeptics' arguments?

The heart of all skeptical arguments against psi phenomena is the Humanist argument that no amount of evidence can ever prove the existence of paranormal events or psi facilities because they are simply impossible. This ad hoc argument is based on the arguments of classical physics which views action at a distance as impossible. As we described above, classical physics assumes that the universe works like a gigantic clock, or billiard balls bouncing around perpetually on a universe size table.

As we have seen in the first four chapters of this book, the triumph of quantum mechanics in 1925 proved that there are no submicroscopic "billiard balls" to bang around. In order to get the billiard ball, you first have to have a conscious observer. Consciousness first, matter second! The laws of physics have been rewritten. Consciousness is a primordial quality of the universe. Matter doesn't create consciousness. Consciousness creates matter! Humanism is an outdated philosophy.

Science, Math and God

A number of physicists including Henry Margenau, David Bohm, and Olivier Costa de Beauregard, have repeatedly noted that nothing in quantum mechanics forbids psi phenomena. Costa de Beauregard, the physicist who suggested the Bell's inequality tests even maintains that the theory of quantum physics virtually demands that psi phenomena exist. Nobel laureate Brian Josephson, who pioneered superconductivity and quantum tunneling has stated that some of the most convincing evidence he has seen for the existence of psi phenomena comes not from the experiments of parapsychologists but rather from experiments in quantum physics. Unfortunately, Johnson was mercilessly criticized for these beliefs by skeptical academics. For more, I refer readers to his home page at the University of Cambridge, Cavendish laboratory.

Contrary to popular belief (and contrary to the statements of skeptical organizations), according to two 1973 surveys consisting of a total of over 1500 respondents, only 3 percent of natural scientists considered ESP to be an impossibility. The only large group of scientists that considered psi to be impossible were psychologists of whom 34% were disbelievers. This disbelief is the result of their unfamiliarity with modern physics and their refusal to look at scientific data proving the existence of psi.

For much more on this subject, as well as exposure to a great deal of modern parapsychological research, please read <u>Science and Psychic Phenomena: The Fall of the House of Skeptics</u> by Chris Carter (Inner Traditions/Bear & Company. Kindle Edition.).

Scientific Theology

Physicists invent theories such as the many worlds hypothesis and the inflationary multiverse at least partly to avoid placing consciousness at the center of the universe and to explain the anthropic nature of the universe we live in. Are there an infinite number of universes? This is a theological question that dwarfs the much simpler question, "Does God exist?"

The complex theories formulated to explain away the baffling implications of quantum theory, relativity and the mystery of why the free parameter fundamental constants just happen to have allowed the creation of an anthropic universe are fascinating and fun to read about. But as we've seen in chapter 3, they rely exclusively on unproven mathematics. They are simply mathematical theologies masquerading as science. None of them have any real-world proof. In the end, they are no more valid than the theologies that postulate the existence of God.

The advantages of belief in Spirit

Considering oneself to be a part of a universe created by a supreme intelligence has at least two major advantages over the materialist belief in a cold, dead universe spawned by random physical processes.

First, **a universe centered on a transcendent intelligence forces us to conclude that our lives have a purpose!** If we owe our existence to a preexisting intelligent agent, then it is safe to assume that we are all here to serve some sort of purpose. The second book in this series is designed to provide real evidence that a spiritual world exists. Once you have looked over the evidence, you will come to see that we are spiritual beings and our purpose is to evolve for as long as it takes to merge with the intelligence that created us. The purpose of this book is to show that modern science no longer prohibits us from crossing this intellectual boundary!

Second, **the belief in God binds our civilization to a moral code** upon which the secular government bases its laws and enforcement mechanisms. This moral code enforces universal honesty that operates over and above the secular law, when no one else is looking. Even non-criminal atheists abide by religious moral laws because they, along with everyone else, absorb them in the course of growing up. This moral code (always based on spiritual teachings) persists as long as the foundational religion of the civilization remains intact.

Granted, some parts of the religious code are simply dogmas which have become meaningless over time, but these relate more to the politics of the age in which these dogmas were first created. Old temporal dogmas are naturally discarded over time, but the core spiritual values always endure. The civilizations that form under a strong moral code based on spiritual insights tend to be stable as long as their core moral codes remain intact.

> Note: The great historian Arnold J. Toynbee believed that societies always die from suicide or murder rather than from natural causes, and nearly always from suicide. He was talking about the late stage decadence that always results from the deterioration of the civilization's moral codes.

With all of the evidence for survival and even the scientific establishment having to admit (however reluctantly) that it doesn't actually have all the answers, there's really no point in arguing with the skeptics. They remain mired in nineteenth century Newtonian physics and ignore the fact that modern physics has undermined their position. It is easier for those who believe in a spiritual world simply to accept that the materialist establishment will never permit their barren philosophy to suffer an indignity as threatening as a psychic foothold for God.

If your belief in the tenants of your native religion has faltered over time because of the relentless assaults on faith constantly showered upon you, you would do well to spend some time exploring the voluminous evidence for survival of the spirit. The evidence in The Structure of Heaven is NOT based on scripture, and none of it is based on religious dogma.

In fact, faith in one's traditional religious belief is fortified by understanding modern science. Always keep in mind that every serious religion is founded on a belief in a spiritual world. This book is an attempt to show just how fine the dividing line between our world and the next actually is. It is simply an added resource that can strengthen your belief in the spiritual world that lies on the other side of that boundary. You really don't need to argue with the rabid materialists that denigrate your beliefs. Science is actually on your side!

Psychic phenomena lie at the foundation of all great religions, and they remain as much a part of our modern lives as they were for the ancients. It's normal to have doubts, but remember that you are not alone, and the evidence makes it clear that a belief in God and a world of Spirit is NOT irrational. It's the skeptics, with their outdated belief in the dogmas of classical physics that are irrational!

The evidence for a spiritual world is really quite extensive. The purpose of this book is to present the reader with an understanding of the limitations inherent in the scientific/materialist philosophy that underlies our modern world, and to excavate a mental landscape that can allow an otherwise materialist mind to accept a spiritual reality. The evidence and rational for a spiritual world can be found in this book's bibliography and its sister volume: "The Structure of Heaven by Martin Spiller.

Bibliography

Adare, Viscount (2011-06-05). Experiences in Spiritualism with D. D. Home. White Crow Books. Kindle Edition.

Atwater, P.M.H. (2010-07-20). I Died Three Times in 1977 - The Complete Story. Cinema of the Mind / Starving Artists Workshop. Kindle Edition

Atwater, P. M. H. (2009-01-09). Near-Death Experiences, The Rest of the Story: What They Teach Us About Living and Dying and Our True Purpose. Hampton Roads Publishing. Kindle Edition.

Alexander III M.D., Eben (2012-10-23). Proof of Heaven. Simon & Schuster, Inc. Kindle Edition.

Allen, Thomas B. Possessed: The True Story of an Exorcism: Iuniverse.com, Inc. 1994

Arcangel, Dianne. Afterlife Encounters: Hampton Roads Publishing, 2005

Barker, Elsa (2011-08-02). Letters from the Afterlife: A Guide to the Other Side, Atria Books/. Kindle Edition. Hatch, David Patterson, 1846-1912 (Spirit)

Betty, Stafford (2014-05-25). Heaven and Hell Unveiled: Updates from the World of Spirit. White Crow Productions Ltd. Kindle Edition.

Barzun, Jaques. From Dawn to Decadence: 500 Years of Western Cultural Life 1500 to the Present: Harper Collins 2002

Bodine, Michael. Growing Up Psychic: Llewellin Publications, 2010

Behe, Michael J. (2001-04-04). Darwin's Black Box: The Biochemical Challenge to Evolution. Free Press. Kindle Edition. ion.

Behe, Michael J. (2007-06-05). The Edge of Evolution . Simon & Schuster, Inc. Kindle Edition.

Berlinski, David (2008-04-01). The Devil's Delusion: Atheism and Its Scientific Pretensions. Random House, Inc... Kindle Edition.

Behe, Michale. The edge of Evolution: The Search for the Limits of Darwinism, Free Press, 2007

Berlinski, David. The Devil's Delusion: Atheism and Its Scientific Pretensions: Basic Books, 2009

Botkin, Allan (2014-05-01). Induced After Death Communication: A Miraculous Therapy for Grief and Loss. Hampton Roads Publishing. Kindle Edition.

Boyce, Mary. Zoroastrians: Their Religious Beliefs and Practices (London: Routledge and Kegan Paul, 1979,

Brittle, Gerald, The Demonologist: The Extraordinary Career of Ed and Lorraine Warren, Graymalkin Media, Los Angeles & New York, 2013

Byrne, Lorna (2009-04-16). Angels in My Hair. Potter/TenSpeed/Harmony. Kindle Edition.

Carter, Chris (2012-02-22). Science and Psychic Phenomena: The Fall of the House of Skeptics. Inner Traditions/Bear & Company. Kindle Edition.

Carter, Chris (2012-08-22). Science and the Afterlife Experience: Evidence for the Immortality of Consciousness. Inner Traditions/Bear & Company. Kindle Edition.

Chalmers, David J. (1996-03-30). The Conscious Mind: In Search of a Fundamental Theory (Philosophy of Mind Series) (Kindle Location 1034). Oxford University Press. Kindle Edition.

Chism, Stephen, The Afterlife of Leslie Stringfellow: A Nineteenth-Century Southern Family's Experiences with Spiritualism, 2005,

Corte, Nicolas. Who Is the Devil? Sophia Institute Press. Kindle Edition.

Choquette, Sonia (2003-07-01). Diary of a Psychic (Kindle Location 3717). Hay House. Kindle Edition.

Crabtree, Adam. Multiple Man: Explorations in Possession and Multiple Personality: 1986

Crumley, Karen (2011-10-01). Growing Up Weird: Confessions of a Closet Medium (Kindle Location 2290). Purple Sage Publishing. Kindle Edition.

Cummins, Geraldine. (2012-12-06). Beyond Human Personality, White Crow Books, Kindle Edition

Cummins, Geraldine (2012-08-06). The Road to Immortality . White Crow Books. Kindle Edition.

Davenport, Reubin Briggs. The Death-Blow to Spiritualism Being the True Story of the Fox Sisters: 1888, [Kindle Edition]

Davies, Paul. The Goldilocks Enigma: Why Is the Universe Just Right for Life?. Houghton Mifflin Harcourt.

Durant, Will (2011-06-07). <u>Our Oriental Heritage: The Story of Civilization, Volume I</u>. Simon & Schuster. Kindle Edition.

Durbin, Deborah (2011-02-25). <u>Oh Great, Now I Can Hear Dead People!</u> Kindle Edition.

Dostoyevsky, Fyodor (2011-02-18). <u>Delphi Complete Works of Fyodor Dostoyevsky</u> (Illustrated) (Kindle Location 41164). Delphi Classics. Kindle Edition.

Emmons. Charles (2003-03-03). <u>Guided by Spirit: A Journey into the Mind of the Medium</u> (Kindle Location 19). iUniverse. Kindle Edition.

Feynman, Richard P., and A. Zee. QED: <u>The Strange Theory of Light and Matter</u>: (Princeton Science Library) (Apr 4, 2006)

Flew, Antony; Varghese, Roy Abraham (2009-10-13). <u>There Is a God</u>. HarperCollins. Kindle Edition.

Flew, Antony; Varghese, Roy Abraham (2009-10-13). There Is a God. HarperCollins. Kindle Edition.

Gallagher, Richard; <u>Demonic Foes; my 25 years as a psychiatrist investigating possessions, diabolic attacks and the paranormal</u>: October 6, 2020; HarperCollins publisher

Gibbon, Edward (2008-03-07). History of the Decline and Fall of the Roman Empire, all six volumes, with active table of contents, improved 2/1/2011). B&R Samizdat Express. Kindle Edition.

Giesemann, Suzanne R. The Priest and the Medium: The Amazing True Story of Psychic Medium B. Anne Gehman and Her Husband, Former Jesuit Priest Wayne Knoll, Ph.D.Kindle Edition.

Goldman, David (2011-09-19). How Civilizations Die: (And Why Islam Is Dying Too). Perseus Books Group. Kindle Edition.

Greaves, Helen (2009-06-25). Testimony of Light: An Extraordinary Message of Life After Death. Penguin Publishing Group. Kindle Edition.

Greene, Brian (2009-01-08). The Elegant Universe: Superstrings, Hidden Dimensions, and the Quest for the Ultimate Theory. Norton. Kindle Edition.

Greene, Brian (2011-01-25). The Hidden Reality: Parallel Universes and the Deep Laws of the Cosmos. Knopf Doubleday Publishing Group. Kindle Edition.

Guggenheim, Bill; Guggenheim, Judy (2012-09-05). <u>Hello from Heaven</u>: A New Field of Research-After-Death Communication Confirms That Life and Love Are Eternal. Random House Publishing Group. Kindle Edition.

Hare, Robert D. <u>Without Conscience: The disturbing world of the Psychopaths among us</u>: Guilford Press, 1993

Hare, Robert D. <u>Snakes in Suits: When Psychopaths Go to Work</u>: Harper, 2006

Hawking, Stephen; Mlodinow, Leonard (2010-09-07). <u>The Grand Design</u>. Random House, Inc. Kindle Edition.

Holland, John with Cindy Pearlman. <u>Born Knowing: A Medium's Journey</u>: Hay House inc., 2003

Holzer, Hans (2004-09-01). <u>Ghosts: True Encounters with the World Beyond</u> (Kindle Location 25472). Black Dog & Leventhal Publishers. Kindle Edition.

Holzer, Hans. <u>Ghosts: True Encounters with the World Beyond</u>: Black Dog & Leventhal Publishers, 2003

Horn, Stacy. Unbelievable: Investigations into Ghosts, Poltergeists, Telepathy, and Other Unseen Phenomena, from the Duke Parapsychology Laboratory (P.S.). Harper Collins, Inc, (2009-03-03)

Ingram, Martin Van Buren (2009-07-12). The Bell Witch Hauntings (An Authenticated History of the Famous Bell Witch: A True Story). Ignacio hills press (TM) IgnacioHillsPress.com and e-Pulp Adventures (TM). Kindle Edition.

Kardec, Alan (Denizard-Hyppolyte-Leon Rivail). The Spirits Book: April 18, 1857

Kelly, Edward F. Irreducible Mind: Toward a Psychology for the 21st Century Rowman & Littlefield Publishers (November 16, 2009)

Kubler-Ross, Elisabeth. On Death and Dying: Simon and Schuster, 1969

Kuttner, Fred; Bruce Rosenblum (2008-10-14). Quantum Enigma: Physics Encounters Consciousness. Oxford University Press. Kindle Edition.

Lanza, Robert and Bob Berman. Biocentrism: How Life and Consciousness are the Keys to Understanding the True Nature of the Universe: Benbella Books, 2009

Lanzara, Joseph; Milton, John (2012-05-30). John Milton's Paradise Lost In Plain English. New Arts Library. Kindle Edition.

T.C.Lethbridge. Ghost and Ghoul: Routledge & Kegan Paul, 1961

Mayhan, Joni (2015-11-27). Ruin of Souls. Kindle Edition.

Matheson, Richard (2007-04-01). What Dreams May Come: A Novel (p. 275). Tom Doherty Associates. Kindle Edition.

Meyer, Stephen. The Signature in the Cell: Harper Collins, 2009

Meyer, Stephen C. (2013-06-18). Darwin's Doubt: The Explosive Origin of Animal Life and the Case for Intelligent Design. HarperCollins. Kindle Edition.

Monahan, Brent. The Bell Witch haunting: St Martin's Press, 1997 (Originally published as Our Family Trouble in 1890 by Richard Williams Bell, the son of John Bell Jr.

Myers, Frederick W. Human Personality and its Survival of bodily Death 1903

Moody, Raymond and Elisabeth Kubler-Ross, Life After Life: The Investigation of a Phenomenon--

Survival of Bodily Death: Harper San Francisco 2001

Moses, William Stanton. Spirit Teachings: Through the Mediumship of William Stainton Moses: RECEIVED DURING THE 1870's

Musser, George. Spooky Action at a Distance: The Phenomenon That Reimagines Space and Time-- and What It Means for Black Holes, the Big Bang, and Theories of Everything. Farrar, Straus and Giroux. Kindle Edition.

Myers, F. W. H. (Frederic William Henry) (2012-01-10). Human Personality and its Survival of Bodily Death. Kindle Edition.

Obley, Carole J. (2010-05-11). I'm Still With You: True Stories of Healing Grief Through Spirit Communication. NBN_Mobi_Kindle. Kindle Edition.

Owen, G. Vale (2012-06-24). The Life Beyond the Veil: A Compilation of Four Classic Books (Life on Other Worlds Series). Square Circles Publishing. Kindle Edition.

Penfield, Wilder, The Mystery of the Mind,

Penrose, Roger (2011-09-06). Cycles of Time: An Extraordinary New View of the Universe. Knopf Doubleday Publishing Group. Kindle Edition.

Carter, Chris (2010-08-23). Science and the Near-Death Experience: How Consciousness Survives Death (Kindle Location 5038). Inner Traditions/Bear & Company. Kindle Edition.

Penrose, Roger and Martin Gardner. The Emperor's New Mind: Oxford University Press, 1989

Puryear, Anne: Stephen Lives! (Scottsdale, Arizona: New Paradigm Press, 1992)

Radin, Dean. The Conscious Universe: The Scientific Truth of Psychic Phenomena: Harper Edge, 1997

Renier, Noreen. A Mind for Murder: The Real-Life Files of a Psychic Investigator, Hampton Roads Publishing Company, 2008

Rhine, J.B., Extra-Sensory Perception After Sixty Years (Rhine, J.B., Pratt, J.G.; Smith, Burke M; Stuart, Charles E; and Greenwood, Joseph A. Extra-Sensory Perception After Sixty Years, Holt: New York, 1940; Humphries: Boston, 1966)

Ring, Kenneth and Sharon Cooper. <u>Mindsight</u>: The William James Center for Consciousness Studies, Institute of Transpersonal Psychology, 2008

Ring, Kenneth. <u>Lessons from the Light</u>:   Moment Point Press Inc.; 1 edition (September 1, 2006)

Ring, Kenneth. <u>Heading Toward Omega</u>: Amazon Digital Services LLC, June 18, 2012

Roberts, Jane, <u>The afterdeath Journal of an American Philosopher—The World View of William James</u>: New Awareness Network, Inc. 1978

Schwartz, Gary E. <u>The G.O.D. Experiments: How Science Is Discovering God In Everything, Including Us</u>: Atria Publishing, 2006

Schwartz, Gary E. <u>The Afterlife Experiments: Breakthrough Scientific Evidence of Life After Death</u>: Atria Books, 2002

Sheldrake, Rupert (2011-04-26). <u>Dogs That Know When Their Owners Are Coming Home</u>: Fully Updated and Revised. Crown/Archetype. Kindle Edition.

Sinclair, Upton. Mental Radio: 1930

Spiller, Martin; The Structure of Heaven; 2020 Amazon.com

Spiller, Martin (2012-06-27). The lower Reaches of Heaven; An exploration of the role of spirits in mental illness. Martin S. Spiller. Kindle Edition.

Stafford, Betty. (2011-06-16). The Afterlife Unveiled: What the Dead are Telling Us about Their World. John Hunt Publishing. Kindle Edition

Stainton Moses, Spirit Teachings through the Mediumship of William Stainton Moses, Free Internet download

Stevenson, Ian. Children Who Remember Previous Lives. McFarland & Company, Inc., Publishers.

Stevenson, Ian. Twenty Cases Suggestive of Reincarnation: University Press of Virginia, 1974

Sugrue, Thomas, There is a River: Holt, Rinehart and Winston, 1942

Swedenborg, Emanuel. Heaven and its Wonders and Hell: 1758 Kindle Edition

Swedenborg, Emanuel (2009-10-04). <u>Angelic Wisdom about Divine Providence</u> (Kindle Location 1). Public Domain Books. Kindle Edition.

Targ, Russell⁣, Harold E. Puthoff and Richard Bach. <u>Mind Reach</u>: Hampton Roads Publishing, (Feb 2005)

Van Dusen, Wilson. <u>The presence of other worlds: The psychological and spiritual findings of Emanuel Swedenborg:</u> reissued by the Swedenborg Foundation 1991.

Van Praagh, James, <u>Ghosts Among Us</u> Uncovering the Truth About the Other Side; HarperCollins e-books

Varghese, Roy Abraham. <u>There Is Life After Death: Compelling Reports from Those Who Have Glimpsed the Afterlife.</u> Kindle Edition

Vasiliev, L.L. and Arthur Hastings PhD, <u>Experiments in Mental Suggestion (Studies in Consciousness)</u>, Hampton Roads Publishing, Sep 1, 2002

Vaughan, Deborah. <u>Talking to Ghosts: Ask Spirits Questions & Get Real Answers with a Digital Voice Recorder:</u> (Electronic Voice Phenomena) Kindle edition

Ward Peter D., and Donald Brownlee, <u>Rare Earth</u>: Copernicus Books, 2000

Warren, Ed; Warren, Lorraine; Lasalandra, Michael; Merenda, Mark (2014-10-03). <u>Satan's Harvest</u> (Ed & Lorraine Warren Book 6) Graymalkin Media. Kindle Edition.

Warren, Ed; Warren, Lorraine; Chase, Robert David (2014-10-03). <u>Graveyard</u> (Ed & Lorraine Warren Book 1). Graymalkin Media. Kindle Edition.

Weisberg, Barbara. <u>Talking to the Dead: Kate and Maggie Fox and the Rise of Spiritualism</u>: Harper Collins, 2004

Wilson, Colin. <u>The Occult: A History</u>: 1971

Wilson, Colin. <u>Beyond the Occult</u>: 1988

Woit, Peter (2006). <u>Not Even Wrong</u>: The Failure of String Theory and the Search for Unity in Physical Law. Basic Books. p. 105

www.ingramcontent.com/pod-product-compliance
Lightning Source LLC
Chambersburg PA
CBHW070536220526
45467CB00003B/963